DARWINIAN POPULATIONS
AND NATURAL SELECTION

DARWINIAN POPULATIONS AND NATURAL SELECTION

PETER GODFREY-SMITH

OXFORD

UNIVERSITY PRESS

OXFORD

UNIVERSITY PRESS

Great Clarendon Street, Oxford OX2 6DP

Oxford University Press is a department of the University of Oxford.
It furthers the University's objective of excellence in research, scholarship,
and education by publishing worldwide in

Oxford New York

Auckland Cape Town Dar es Salaam Hong Kong Karachi
Kuala Lumpur Madrid Melbourne Mexico City Nairobi
New Delhi Shanghai Taipei Toronto

With offices in

Argentina Austria Brazil Chile Czech Republic France Greece
Guatemala Hungary Italy Japan Poland Portugal Singapore
South Korea Switzerland Thailand Turkey Ukraine Vietnam

Oxford is a registered trade mark of Oxford University Press
in the UK and in certain other countries

Published in the United States
by Oxford University Press Inc., New York

British Library Cataloguing in Publication Data

Data available

Library of Congress Cataloging in Publication Data

Data available

Typeset by SPI Publisher Services, Pondicherry, India
Printed in Great Britain
on acid-free paper by
MPG Books Group, Bodmin and King's Lynn

ISBN 978–0–19–955204–7 (Hbk.)
978–0–19–959627–0 (Pbk.)

1 3 5 7 9 10 8 6 4 2

For Jane

PREFACE

This book is about evolutionary theory—Darwin's theory, as modified by his many intellectual descendants. It is primarily about evolution by natural selection, the process by which populations change through a dynamic of variation, inheritance, and reproduction. But natural selection is discussed in the context of a more general Darwinian view of life, and is seen through the lens of philosophy of science.

Believing that its topic is important to philosophers, biologists, and people outside both those categories, I have tried to write the book for all three kinds of reader. This has been done by organizing the book in several layers. The eight main chapters will, I hope, be accessible to people with very little background in philosophy or evolutionary biology. With the exception of some footnotes, they contain almost nothing technical. I have also tried to avoid jargon as much as possible (adding some extra explanations of terms that really matter). Within those eight chapters, the first five function as a unit. They develop and defend my view of natural selection. This view is, as my title suggests, organized around the idea of a "Darwinian population." A Darwinian population is an extraordinary arrangement of ordinary things. Its components are startlingly routine—births, matings, deaths, heredity—but its products can be very far from routine. I look at what the process of evolution by natural selection requires, and at what it can explain. I also try to describe how the world appears when we see Darwinian populations as one of its key elements.

That picture is summarized at the end of Chapter 5. The next three chapters look at more detailed topics and debates. These include the "gene's eye view" of evolution and the idea that cultural change is itself a Darwinian process.

The footnotes add a second layer to the book. They contain further connections to the literature in both philosophy and biology, sketches of models, clarifications and defenses, and comments about extra paths that can be followed. More jargon has been allowed in to keep them brief.

Thirdly there is an Appendix, which contains technical ideas relevant to the main chapters. So most of the Appendix supplements arguments given earlier. The exception is the last section, which is free-standing and presents a different way of representing a large range of Darwinian and non-Darwinian phenomena. The Appendix can be read in pieces, after relevant chapters, or as a unit.

I have a large number of people to thank, especially the many who wrote detailed comments on an entire earlier draft. These were Richard Francis, David Haig, David Hull, Ben Kerr, Arnon Levy, Elisabeth Lloyd, John Matthewson, Samir Okasha, Kim Sterelny, and Kritika Yegnashankaran. Every one of them was responsible for crucial improvements. Each new set of comments, in fact, seemed to turn yet another part of the book upside-down, also sending waves through the rest. Haig and Sterelny were responsible for particularly high seas.

With a group of commentators like that, it seems that every argument in the book should now be completely watertight. I am sure that is not so, but whatever its remaining flaws, the book has benefited enormously from having such a diverse and knowledgeable group pay close attention to its development. The work also owes much to the longer-term influence of collaboration with Ben Kerr, and the miraculous clarity and originality of his thought. Sections of the Appendix draw directly on work with him, but his influence extends throughout.

For generous help with biological matters, I am grateful to Rick Michod, Katherine Preston, Sally Otto, Bob Cooke, Jacques Dumais, Armin Hinterwirth, Eva Jablonka, Marshall Horwitz, and Karola Stotz. Ellen Clark sent acutely useful comments on the middle chapters. Drew Schroeder pulled apart early versions of the spatial representations. Discussions with Glenn Adelson led, just before the actual writing, to a reorientation of much of the argument. Jane Sheldon made innumerable improvements to both the content and style, and also found the ideal cover image. I was helped by additional discussions and correspondence with Dick Lewontin, Jura Pintar, Dan Dennett, Thomas Pradeu, Laurie Paul, Patrick Forber, Lukas Rieppel, Paul Griffiths, Brett Calcott, and Justin Fisher. Eliza Jewett skillfully rendered the figures. I am grateful to Harvard University for an exceptional intellectual environment, along with a well-timed sabbatical. Working with Peter Momtchiloff of Oxford made it very clear why he has an excellent reputation as an editor.

PGS
May 2008

CONTENTS

1. Introduction and Overview   1
2. Natural Selection and its Representation   17
3. Variation, Selection, and Origins   41
4. Reproduction and Individuality   69
5. Bottlenecks, Germ Lines, and Queen Bees   87
6. Levels and Transitions   109
7. The Gene's Eye View   129
8. Cultural Evolution   147

   Appendix: Models   165
   Bibliography   185
   Index   203

CHAPTER 1

INTRODUCTION AND OVERVIEW

1.1. Science, Philosophy of Science, Philosophy of Nature

[*Organization of evolutionary biology; foundational discussions of natural selection; science, philosophy of science, and philosophy of nature.*]

This book is about evolution by natural selection—about the process itself, and our attempts to understand it through scientific theorizing. Each topic illuminates the other.

Evolutionary biology as a whole can be seen as organized around two central collections of ideas. One is summarized by the "tree of life." This is the hypothesis that all organisms on earth are related to each other by common ancestry, and if we "zoom out" of a chart of this total set of relations of ancestry and descent, the genealogical relationships between species form the rough shape of a tree. The second is our account of how change occurs within populations or species. This is where we find, among other things, evolution by natural selection. Variations arise within populations, in a haphazard and undirected way. Some of these variant characteristics lead the individuals who bear them to have more offspring than others. When these favored characteristics are inherited across generations, the population will change.

The history of attempts to describe the essential features of this process is now a long one. Darwin's descriptions in the *Origin of Species* (1859) were mostly fairly concrete. That is, they were aimed at capturing how the process works in actual-world organisms and environments. Quite quickly, though, there were moves towards a more abstract treatment of his central ideas, and this tradition has continued. It has seemed to many that Darwin identified a general pattern, a sort of schematic machine, that might be found in many domains and is not dependent on the contingent features of biological systems here on earth. This tradition includes what I will call the "classical" approach to the abstract description of Darwinism, in which variation, heredity, and differences in reproductive output are seen as comprising a recipe for change in a wide range of systems. More recently, the tradition includes descriptions of evolution in terms of "replicators," faithfully copied structures that endure for long stretches of time and build the rest of the biological world around them.

The reasons behind the attempt to give an abstract description of this kind are roughly twofold. One is the search for a pure understanding of the evolutionary process itself. The other is the search for a theoretical tool that might have applications in new domains. This book is a continuation of that project. There is now a substantial history of such descriptions, as noted above, but I hope to improve on earlier ones and also take the project further. This is to be done partly through the continuation of existing lines of discussion and partly via input from new sources. Those include looking at the problem through the lens of recent philosophy of science.

The book is intended to be simultaneously philosophical and biological. Its topics are approached from the viewpoint of philosophy, but the arguments are intended to have consequences for biology as well. To describe the mix, I will distinguish three kinds of projects and subject matter: science, philosophy of science, and what can be called "philosophy of nature."

The focus of science is the natural world. Science investigates the world not with a rule-governed "method," but with something more like a *strategy*. Ideas about how the world works are developed and assessed in a way designed to make those ideas both internally coherent and responsive to observation. One part of this process involves exploration of the "inner logic" of theoretical ideas—their resources, their powers of explanation in principle, how they connect to other pieces of theory, and what sort of data would support or tell against them.

The focus of the philosophy of science is science itself, the process described just above. The aim is to understand how science works and what it achieves. Here we ask what kinds of contact with the world theories can have—how they function as representations, how they can yield understanding. We ask about the role of worthy but vexed goals such as truth, simplicity, and explanatory power, and about the nature of evidence, testing, and scientific change. Such work can cast its net widely, to capture all of science, or narrowly, to comprehend a small part of it, such as evolutionary biology.

Work of this kind feeds back usefully into science to the extent that it is good for science to be self-conscious. Not everything is best done self-consciously, and it is not obvious whether science is. In Thomas Kuhn's account of scientific change, for example, science works best when the "normal scientist" is somewhat mistaken about what he or she is really doing, and about how the large-scale course of science runs (Kuhn 1962). This is an intriguing and slightly macabre view, overstated at the least. But Kuhn's picture, which is certainly coherent, indicates that it is not obvious and inevitable that philosophy of science should be helpful to science. And aside from any such feedback, the philosophical understanding of science is a goal in itself.

Thirdly, there is a different kind of philosophical work. Now the focus is on the natural world again. But the focus is on the natural world as seen through the instrument of science. This is the project of taking science as developed by scientists, and working out what its real message is, especially for larger questions about our place in nature. So we aim to use scientific work to inform our view of the world, but we do not determine this view using science in its "raw" form. Instead we take the raw science on a given topic and work out, philosophically, what exactly the work is saying. Reviving an old term, this project can be called "philosophy of nature."

Some might wonder whether there is really work of this kind to be done. If we are scientifically minded, why don't we take our view of the world from science in its raw form? If, on the other hand, we don't trust what the scientists are saying, shouldn't we look to other sources altogether?

A person's attitude to this third project will depend on how they think science operates—it will depend on their views in the philosophy of science. Here is one package of ideas within which this third project makes good sense. Science is an unusually powerful tool for investigating what the world is like. But ideas are developed within scientific communities according to the demands of science itself. The results include concepts which have contours that fit the practicalities of scientific work—the demand for questions to be tractable, for work to be cooperative (also competitive), for contrasts between options to be usably sharp. We also encounter language that is infused with subtle—almost invisible—metaphors, categories that are shaped by the toolkit as well as the phenomena, and simplifications that oil the wheels of day-to-day work. When we export a picture of the world from the immediate context of science into a broader discussion, the features of scientific description that have their origin in these practicalities become potentially misleading. The "broader discussion" here might be overtly philosophical (within ethics or philosophy of mind, for example), or it might be even broader, and less academic, than that. But scientific information generally needs processing before it feeds into discussions of those kinds. Work of this sort will also often aim at synthesizing the results of a number of different scientific fields, working out how they fit together—or fail to fit—into a coherent package.

So philosophy of nature refines, clarifies, and makes explicit the picture that science is giving us of the natural world and our place in it. Calling it "philosophy" does not mean that only philosophers can do it. Many scientists, including many discussed in this book, undertake this kind of work. But it is a different activity from science itself.

This book moves constantly between these three kinds of investigation. The starting point is philosophy of science—a look at evolutionary theory from a philosopher's point of view. I examine how evolutionary biology attempts to represent and understand a certain set of natural phenomena. This sort of work, I said, need not always feed back productively into the science, but here my hope is that it may do so. There are issues that become clearer by attending to the roles of different kinds of theoretical "vehicles" used in this area, and I argue for new views about the relations between some central evolutionary concepts. The resulting picture is also used to induce a reflective attitude towards our psychological habits, when we think about the biological world. Those habits, which affect how we do both science and philosophy, are the upshot of the relations between how our Darwinian world runs and how our evolved minds operate.

1.2. A Statement of Central Themes

[*Recipes and replicators; the minimal concept; marginal cases and paradigms; spatial tools; levels and transitions.*]

The existing literature contains two main traditions of abstract description of natural selection. One I will call the "classical" approach. This is a collection of summaries that have roughly the following form: evolution by natural selection results whenever there is a population in which there is variation between individuals, which leads to those individuals having different numbers of offspring, and which is heritable to some extent. These formulations are often expressed as a kind of recipe: if we have those ingredients, evolutionary change will follow. The recipe is applicable, in principle, to any entities capable of some form of reproduction.

This "classical" approach is the starting point for my own analysis, but the standard formulations have a number of problems. There are cases where the ingredients are present but change does not occur, and cases where change of (seemingly) the right kind occurs without the ingredients being present. Sometimes the content of a summary seems at odds with the commentary the author gives about it.

These problems arise for philosophically interesting reasons. Standard recipes for change by natural selection are products of trade-offs, often unacknowledged, between the desire to capture all genuine cases of natural selection in a summary description, and the desire to describe a simple and causally transparent machine. These two motives correspond to two different kinds of understanding that we can seek in any theoretical investigation. One approach is to try to give a literally true description of all cases of the phenomenon we are interested in. Another option is to directly describe one class of cases, which are simple and tractable, and use these as the basis for a more indirect understanding of the others. Understanding is achieved via similarity relations between the simple cases we have picked apart in detail, and the cloud of more complicated ones.

The "simple" cases in such an exercise might be a simpler set of real-world cases, or they might be a set of fictional cases that we arrive at via imagined modification of the real ones. Either way, in this second approach understanding is achieved indirectly, by means of models, rather than by direct description of the mass of empirical phenomena. Unacknowledged movement between these two approaches is operating in many foundational discussions of natural selection. One might reply at this point: surely we do not need to choose between these two approaches, but should often apply both? That is indeed what I do here, but doing this requires setting things up in a new way.

The second tradition of abstract description of natural selection is more recent; this is the "replicator" approach, developed originally by Richard Dawkins (1976) and David Hull (1980). A replicator, roughly, is anything that makes copies of itself, or induces copies to be made. The general picture we are offered is something like this. The first replicators arise at the origin of life itself, and are no more than single molecules. There is a raw evolutionary competition among these simple replicators; those that replicate faster and more accurately, and remain intact for longer, become more common. But eventually they engage in large-scale activities of cooperation—when advantageous to them—which include the building of "vehicles" or "interactors." These are larger units that house replicators and (especially in Dawkins' account) serve the replicators' purposes. Modern replicators here on earth are genes, and (perhaps, recently) some culturally transmitted entities as well.

According to this view, replicators are essential ingredients in any Darwinian process. The inner logic of evolutionary theory is organized primarily around the idea that replicators will evolve to further their replication, and other living structures exist as products of replicator action. We are to understand the living world by treating each gene or other replicator as "the center of a web of radiating power" (Dawkins 1982a: vii).

I will be much more critical of the replicator view. It is not true that evolution by natural selection requires replicators. Part of the appeal of this framework has to do with the fact that the replicator view is designed to mesh with a particular way of thinking about evolution, which I will call "agential." Here we think of evolution in terms of a contest between entities with agendas, goals, and strategies. I see the agential view of evolution as something of a trap. It has real heuristic power in some contexts, but also has a strong tendency to steer us wrongly, especially when thinking about foundational issues. And once we start thinking in terms of little agents with agendas—even in an avowedly metaphorical spirit—it can be hard to stop.

Some people might wonder at this point whether we *need* an abstract verbal description of natural selection that covers all cases. Perhaps words are not up to the task. When evolutionary theory was young, it was possible for Darwin to express all his principles in simple (and elegant) English. But one trend in evolutionary theory over the last hundred years has been towards formalization. Rather than a verbal description, perhaps we should be looking for a master equation, an $E = mc^2$ or $F = MA$ of evolution. Words, it might be said, become blunt instruments once a scientific theory reaches the state of modern neo-Darwinism.

Although most of this book is written very informally, the discussion will often have an eye on the formal models. I argue that these models, which are packed with idealizations, serve a different role from that of the verbal descriptions, and do not replace them. Most discussion of the "master equation" approach to evolution will be in an Appendix, though, to keep the book as non-technical as possible.

The framework used in the rest of the book is introduced in Chapter 2. The central concept, as the title suggests, is that of a "Darwinian population." This is a population—a collection of particular things—that has the capacity to undergo evolution by natural selection. A "Darwinian individual" is any member of such a population. Darwinian individuals will be discussed quite a lot, but the idea is that the population-level concept comes first. The first main step is description of what I will call the "minimal concept" of a Darwinian population, and a corresponding category of change. The minimal concept features three ingredients that are familiar from the classical approach: variation in individual character, which affects reproductive output, and which is heritable. The minimal concept functions as something of a stepping stone, however, and its role recedes as the full framework is laid out.

The second step is a kind of fragmentation, recognition of a family of different kinds of Darwinian processes. The minimal concept is very permissive, including both biologically significant cases and many that seem almost trivial. If one derives a picture of natural selection solely from the minimal concept, it can seem hard to understand how Darwinism could be so important. But the process described in the minimal concept should not be installed as "the" Darwinian process. Significant Darwinian processes have extra features, and these can in some cases be described abstractly. So within the area staked out by the minimal concept, we can identify a category of *paradigm* Darwinian populations. This is the kind of system that can produce novel and complex organisms, highly adapted to their circumstances.

At the other end of things, we can also identify a category of *marginal* Darwinian populations. This is not a category of cases within the boundaries of the minimal concept; rather, these are populations that do not clearly satisfy the minimal requirements, but do approximate them. They have a partially Darwinian character.

So we have, at this stage, a *minimal* concept (broad and permissive), a concept of a *paradigm* case (much narrower) and a concept of a *marginal* case (merely approximating the minimal criteria). But it is clear that "paradigm" status is not a discrete category, and there are various ways—various dimensions—via which a Darwinian population can be a more significant one as opposed to a more trivial one. I will introduce a spatial framework to represent some of these possibilities. This part of the argument is more speculative, but the aim is to show that the paradigm cases occupy one region of the space, and the marginal cases another. The minimal criteria pick out a large region, including the paradigms, and this region has vague borders, blending first into marginal cases and then into cases that have no Darwinian characteristics at all.

Up to that point in the argument a key concept is taken for granted: reproduction. Reproduction is at the heart of Darwinism, but it is surrounded by puzzles. The clearest and most familiar cases of reproduction involve the sexual generation of a new individual who is biologically similar to its parents but genetically distinct from them, and whose life develops from a one-cell stage. But there are lots of unclear cases. If a plant sends off a runner, which leads to a new structure that resembles the old one and can live independently, is that reproduction or just growth of the old individual? There are also problems posed by "collective" entities, such as social groups and symbiotic associations. Do bee colonies reproduce, as well as bees? Do buffalo herds reproduce, as well as buffalo? Do tightly linked symbiotic entities like lichens comprise Darwinian individuals of their own? Is the reproducing entity the part or the whole?

That question should not be seen as an either–or; both may reproduce, and hence participate in Darwinian processes at their own level. In the middle chapters I treat reproduction itself in a gradient way, and again employ spatial tools to impose order on the unruly variety of cases. The aim is to defend a view like this: "reproduction" comes in clearer and more marginal forms, sometimes being poorly distinguished from growth, and sometimes being unclear because the reproducing entities themselves have questionable status as individuals. One way that a Darwinian population can be a marginal one is by having individuals linked by reproductive relationships that are themselves marginal. "Marginal" cases of reproduction are not those that *look* strange, given what we are familiar with and given the everyday meaning of the term. Marginal cases of reproduction have a different evolutionary role from more definite ones.

This gives us a new way of addressing some old puzzles—partly biological, partly philosophical—about the "levels" or "units" of selection. Many selection processes seem to be potentially describable at more than one level, and it can be difficult to work out whether and how to choose between these descriptions. For example, when (if ever) is there evolution in which social groups of organisms are the units selected, as opposed to the individuals of which the groups are composed? If the answer is "never," then should selection on whole organisms be seen as really cases of selection on individual genes? When we have biological objects arranged in a hierarchy of parts and wholes, how do we decide the level in the hierarchy at which selection is acting?

My approach to all these questions is to ask separately, at each level, how well the entities at that level meet the criteria for being a Darwinian population. If a system contains genes, chromosomes, cells, organisms, and groups, for example, the framework is applied in a uniform way to all of these. This might seem like the obvious thing to do, but it has been quite common to do something different. I will discuss why that has happened. I then argue that the ideas in earlier chapters enable us to handle some long-standing puzzles over levels in a new way. What we must grapple with when addressing many of these questions is the role of marginal and partial cases of Darwinian populations, and the gradations between these and the paradigm cases.

The spatial tools developed in the middle chapters have two roles. One, emphasized above, is providing a compact way of categorizing different cases. Darwinian populations differ in a host of respects, and one way to impose order is to pick some parameters that can be numerically scored in at least a rough way, and represent populations in an abstract space where each feature corresponds to one dimension of the space. But the spatial representations also have a more dynamic application. When a population evolves it not only changes the characteristics of its constituent organisms. It also changes, as a consequence, *how* it evolves, the manner in which further change will come about. Organisms can evolve higher-fidelity reproduction, or lower; tighter integration of the population, or looser; clearer or vaguer borderlines between generations. The parent–offspring relation itself can be sharpened, obscured, transformed, or lost altogether. So as well as understanding populations as occupying points in a Darwinian space, we can represent some kinds of change as movement through such a space.

In addition to movement of a single population, there is the creation of new Darwinian populations from old ones. This gives us a way of thinking about many of the "major transitions" in evolution, especially those in which lower-level entities associate, cooperate, and eventually form a population of higher-level individuals. The two most conspicuous transitions of this kind are the evolution of the eukaryotic cell from associations (of a swallowing kind) between various prokaryotic (bacteria-like) cells, and the evolution of multicellularity from single-celled life.

The framework developed here is intended to be useful when thinking about such transitions, especially their intermediate stages. Then we often find populations that have a marginal Darwinian status—for example, collective entities that are somewhat organism-like, but not all the way there. A great array of these can also be seen in aquatic life, in which there is a variety of forms of *partial* integration of cells and simple organisms into collective entities, including seaweeds, corals, and sponges.

As a transition of this kind occurs, a population may appear first as a marginal case, from a Darwinian point of view—an ensemble of collective entities who can only be said to reproduce in an extended, generous sense, who barely count as individuals at all. But there may be successive increases in integration, until the entities display a well-defined mode of reproduction at the higher level, with heritable variation in the traits found at that level. The collective becomes a paradigm case. And as this happens, the original lower-level population tends to be pushed *away* from paradigm status. It is a familiar theme that successful maintenance of higher-level organization requires that competition between lower-level entities (cells, for example) must be suppressed. The suppression of this competition works by the curtailing of the independent evolutionary activities of the lower-level entities. One Darwinian population can "de-Darwinize" others.

In the last chapters I look more closely at some controversial topics: the "gene's eye view" of biological evolution and the Darwinian perspective on cultural change. In both cases I proceed by applying, again directly and uniformly, the framework defended earlier. This means that genes are treated in a materialist way, as small parts of organisms, with special causal properties and the capacity to be reliably replicated. They comprise very unusual Darwinian individuals, entities whose reproductive activities and status *as* Darwinian individuals are dependent on cell-level and organism-level activities, especially sex. I resist the idea that genes are the basic units of Darwinian evolution in some strict or ultimate sense, but also discuss some special phenomena where a gene's eye view seems the key to understanding.

I then turn to cultural change. The temptation to treat culture in Darwinian terms is almost as old as Darwinism in biology. Thinking about cultural change in Darwinian terms requires further exploration and stretching of the concept of reproduction. Though many have tried to force cultural phenomena *squarely* into a Darwinian framework, this is another domain populated by partial and marginal cases. There are some features of cultural change that are Darwinian, and much that is not. I try to say where the division lies. Most of the Appendix, which concludes the book, is a supplement to earlier chapters. The exception is the last section, which formalizes some parts of the overall picture and is free-standing.

1.3. Population Thinking and the Pull of Agents and Essences

[*Population thinking; from which we are easily diverted; folk biology's model of organisms; psychology of explanation; evolutionary processes as evolutionary products.*]

The previous section summarized the main positive themes of the book. This section introduces some of the more critical lines of argument, and their connections to a larger picture in which the treatment of natural selection is embedded.

I will begin by saying more about one of my criticisms of the replicator approach. In the previous section I said that this approach is, in many of its presentations, designed to mesh with an "agential" way of thinking about evolution. Evolution is treated as a contest between entities that have purposes, strategies, and agendas. This sort of description can be applied to many biological entities; organisms, for example, might be said to "battle to increase the representation of their genes in future generations." But it is often now applied to genes and other replicators themselves. Replicators act to further their own replication.

The agential perspective on evolution has always been an uneasy mix of the metaphorical and the literal. Different forms of this way of thinking will be discussed later, but all talk of benefits and agendas comes with a peculiar psychological power. This mode of thinking engages a particular set of concepts and habits: our cognitive tools for navigating the social world. David Haig, a biologist who is enthusiastic about the "strategic" approach to genes and evolution, argues that this is, literally, a way for us to be smart when we think about evolutionary problems. The social-and-strategic domain is one where our minds are powerful, as evolutionary psychologists have argued (Cosmides and Tooby 1992). When we think about agents and agendas, we think differently and more acutely than we do about abstract logical and causal relations. The strategic perspective on evolution is a way of scientifically engaging this high-powered side of our minds.

I agree with Haig that we think in distinctive ways when we apply an agential framework, but don't think this is such a good thing. Especially when we think foundationally about evolution, the agential perspective can be seen to steer us wrongly. And once taken up or switched on, these are psychological tools that are hard to put down; they have a compelling, almost addictive, narrative appeal, and tend to send us down specific paths. This claim, again, applies to all agential views of evolution, not just those invoking replicators. But the introduction of small and hidden agents has a particular power. It can lead to an acute form of what Richard Francis (2004) has—dramatically but accurately—called "Darwinian paranoia." Darwinian paranoia is the tendency to think of all evolutionary outcomes in terms of reasons, plots, and strategies. An agenda is a powerful explainer. Once introduced to the possibility of understanding a phenomenon in terms of a grand rationale, we become reluctant to settle for less. One agenda might be exchanged for another, but this becomes the *kind* of understanding we are after. The application of an agenda to the empirical facts might be indirect and constrained, but an agenda "makes sense" of things for us in a way that no mere catalog of efficient causes can.

The picture of agential thinking I employ here can be contrasted with one developed by Robert Wilson, in a related context (2005). For Wilson, any causally active and physically bounded individual is an "agent"; a carbon atom or a brick can qualify, as well as a gene or an organism. And this concept can be used, for Wilson, to analyze how causal description works quite generally in science. Wilson also suggests there is a hidden role for "cognitive" metaphors in *all* of this large category of causal thinking. This applies even to causal description of inanimate things, though presumably it has a less powerful role there. In contrast, the picture I apply understands "agential" thinking in a stronger way than Wilson, and sees it as more specific to particular domains. It is not part and parcel of all causal talk, and I argue that it is a source of trouble in evolutionary thinking.

So one theme of this book is how some intuitive habits of categorization and explanation relate to the picture of the world that Darwinian theory gives us. The core Darwinian concepts have been developed to enable us to deal scientifically with a particular domain. But there are older ways of thinking about this domain that exert a strong influence—in informal contexts, in philosophy, and in biology.

In a classic discussion, Ernst Mayr (1976) claimed that one of Darwin's pivotal achievements was to "replace typological thinking with population thinking." Population thinking is a controversial and slippery idea (Sober 1980, Lewens 2007, Ariew 2008), but I think there is something very important in Mayr's argument.

Much of Mayr's discussion focused on the role of variation within species. Earlier "typological" views, Mayr said, tended to see variation within species as imperfection in the worldly realization of ideal "types." He traced these views back to Plato. Population thinking, in contrast, treats "types" as mere conceptual tools, treats populations themselves as the basis for groupings and kinds, and attends to the causal importance of the variation within them.

This emphasis on individual variation is not a defeatist step away from theorizing ("everything is unique, so general theories always fail"), but a recasting of the subject matter of biology in accordance with better theories about how biological systems actually work. A population is a physical object, bound by ancestry and other causal relations, internally variable at any time and changing over time. To the extent that organisms fall into well-marked and recognizable "kinds" that we can give straightforward species names to, this is a contingent consequence of populational processes. A well-marked kind can split or dissolve, starting tomorrow, if local conditions push it that way.

Population thinking, in roughly Mayr's sense, is indeed how we succeeded in getting a handle on the large-scale structure of the living world. But it is something from which we are easily diverted. It is a latecomer in the development of thought, and not merely for local historical reasons. Mayr said that Darwin replaced typological with population thinking, but it would be more accurate to say Darwin *showed how* to replace it. Typological thinking is tenacious, and this is a battle that must be constantly re-fought. Population thinking goes against a set of psychological habits, a conceptual toolkit, that we naturally tend to use when dealing with the living world.

In making this claim I use elements from several strands of recent empirical psychology, especially work on "folkbiology" (Medin and Atran 1999, Griffiths 2002), combined with more familiar observations from the history of ideas. This old conceptual toolkit comprises at least three elements, sometimes explicitly defended and sometimes operating in a more implicit way.

The first is an *essentialist causal model of organisms*. According to this model, each organism has a type or underlying nature. That nature is *expressed* in its observable features. The roles of underlying nature and contingency of expression often take on a normative loading; when the inner nature is not faithfully expressed, something has gone wrong.

The second element is what Sober (1980) has called a *natural state model* of population variation and change. Species are taken to be collections of individuals that express a type. Perturbation of the population due to happenstance variation leads to a restoring force or tendency, pushing back towards better realization of the type. From a Darwinian point of view, in contrast, the current genetic and phenotypic composition of the population is always the basis for further variation and change. If the population has been pushed by local causes away from an earlier state, there is no intrinsic tendency to return to the previous form. This makes the evolution of a species open-ended, not bounded by any "fixed sphere of possibility."

The third element is a *teleological outlook on biological activity*. This connects to the agential framework discussed earlier, but is broader.

Teleology has been a huge topic in the philosophy of biology; here I mean the status of our tendency to treat biological activities in terms of purposes, goals, and proper functions. It is usually assumed that the intentions of an intelligent designer or user of an object can be the basis for teleological description in a straightforward way. The problem is whether and how these terms can be used in the absence of this overt role for intelligence. In Aristotle, a teleological mode of thinking was the basis for a complete treatment of the natural world. That view was largely supplanted by the "mechanical" philosophy during the scientific revolution, at least when applied to the physical domain. Understandably, these ideas have endured for longer in biology. They have exhibited an uncertain relation to the Darwinian point of view. Sometimes Darwinism is seen as demolishing the last elements of a teleological outlook, but at other times Darwinism is seen as constructively domesticating these ideas, showing they have a limited but real application to biological processes.

The less aggressive of these attitudes was often seen in late twentieth-century philosophy of biology (Buller 1999). A thin form of teleological description can be grounded in a Darwinian view. For example, the Darwinian can say that the function of a body part is the thing it does that has led to its being favored by natural selection. In that thin sense, the function is what that structure is "supposed" to do. This is a very deflationary sense of "supposed to." Any other talk of purposes and goals, except where it is based in the intentions of some intelligent designer or user, is regarded within this Darwinian view as merely metaphorical.

Recently, some philosophers have wanted to restore the teleological outlook on biological activity to a more significant and robust status (eg., Thompson 1995). Often this is based on the idea that a teleological way of looking at things yields a special kind of understanding. Via the use of normatively loaded concepts of purpose and nature, we are able to "make sense" of natural phenomena. I agree that the *intuitions* these philosophers work with are real; we have distinctive habits of categorization and explanation that we find ourselves applying to the living world, and these include essentialist and teleological forms of understanding. But these intuitions are part of a set of habits and ideas that steered us wrong for centuries, as far as theorizing is concerned, and had to be overcome to develop the Darwinian view. The *feeling* that some particular way of looking at things yields understanding should not always be taken at face value, is not the end of the matter.

So part of the background to the critical side of this book is the idea that there are entrenched aspects of our psychology that continue to affect our thinking about the biological world, even when doing science and philosophy. When I talk of "entrenched aspects of our psychology," I refer to *some* mixture of a typical and cross-cultural human psychological profile, plus an overlay deriving from contingent features of the history of Western thought. This is no surprise: the biological domain is one with which we have a long history of pressingly immediate practical relations. It is also one that adjoins and overlaps the social domain where folk-psychological concepts of intention and purpose are most straightforwardly applicable, and it is sufficiently complex to put a premium on compact schemata and models with which we can impose order on it. Correspondingly, an additional aim of this book is to extend and re-assert the power of the perspective on the living world that Mayr gave partial expression to with his concept of population thinking. This includes extensions that Mayr himself, and others in his tradition such as David Hull, might not accept. I will also, towards the end of the book, look at where population thinking breaks down; a naturalistic view does not involve treating *all* collections as populations. But often in the chapters to follow I will emphasize its strengths. And to finish this chapter I will fill out this perspective in more detail, accentuating contrasts with the older set of habits described above.

Let us go back to Mayr's point about the contrast between populational and typological ways of thinking about variation within populations and species. From a typological viewpoint, this variation reflects imperfection in the manifestation of a type by the messy empirical world. Variation *between* species is the basis for categorization and explanation; variation *within* them reflects imperfection in the worldly realization of forms. But as Lewontin has put it, Darwinian evolution is a machine that turns the latter kind of variation into the former. All the striking differences between kinds of organisms, and the traits that make particular species so distinctive, have their origins in variation within populations, filtered and magnified in a way that yields large-scale change.

The Darwinian machine referred to here is the package of factors producing change within populations, combined with the fragmentation and splitting of populations. Via splitting, a population becomes two things with distinct evolutionary paths, accumulating their own responses to accident and local circumstances. This yields an ensemble of populations. Suppose we then "zoom out," so the internal structure of each population disappears, and we also see these objects as extended in time. That generates the other central Darwinian idea distinguished in the first section of this chapter: the tree of life. This is the idea of a network of ancestry and descent linking all organisms—all individuals as well as all species—going back to a single root.

The fact of a single root is contingent, and the assertion of a tree structure is an approximation, especially with respect to life's early stages. But once some such structure is discovered and described, it becomes an enormously powerful idea. Suppose you are looking at a hillside. You see an oak, a cypress, an ant, a bird, a frog, and a stretch of grass. You can immediately infer common ancestry facts for all the living things you see: the oak and the grass having a nearer common ancestor than either has with the cypress; the bird and frog having common ancestors before either shares one with the ant; and eventually a common ancestor, a billion or more years ago, for the oak, the frog, and yourself. The organisms form lineages stretching back in this way, and so do the cells within them, each cell arising from a cell division (or a sexual fusion of cells), and every cell membrane arising from an earlier membrane, via a continuous chain of growth and division.

People sometimes talk of the tree of life as a mere metaphor, but that is selling the idea too cheaply. Representations of life as an approximate tree can be treated as abstract representations of real facts, as subway maps are. Like a subway map, a tree representation is not merely a useful tool, but one that works *via* an abstract correspondence with real things that it is intended to represent. This representation is a partial one; there are anomalies, cross-links, and exceptions. Life is only roughly a tree, but a great deal follows from its being roughly a tree.

The focus of this book is evolution within populations. But evolution within populations is viewed in the context of the tree. The mindset that results can be illustrated by looking again at the message from population thinking regarding species and other biological categories.

The obvious surprise regarding biological kinds that we get from evolutionary theory is that species themselves are not fixed types. But another surprise is that there is variation with respect to the *kinds of* similarity and division that are found. Some organisms, like humans and chimps, fall into conspicuous and (presently) sharply bordered kinds. That fact—what sort of collection is formed, not just what the organisms in the collection are like—is a consequence of local Darwinian factors. Other organisms do not form kinds in this sense. In oaks, species barriers dissolve via complicated webs of hybridization—botanists talk of "hybrid swarms" of oaks. In several kinds of organisms there are "ring species," in which a chain of locally similar and interbreeding organisms gives rise to far ends of the chain that are markedly different and do not interbreed. Asexual organisms are different again, forming individual lineages that have, in principle, an isolated quality, but which nonetheless form something *like* species kinds—hence the term "quasi-species." So as we move around different parts of the tree of life, we find not only different-looking organisms, but different-looking kinds as well, each formed by its own history, ecology, and genetics. There is variation and evolution of kind-hood, as well as of the organisms themselves (Dupré 2002, Wilkins 2003).

The same way of thinking can be brought to bear on evolution by natural selection itself. Evolutionary processes, and the differences between them, can be abstractly characterized. But the forms taken by these processes on earth are consequences of the organisms (and the environments) that evolution has generated. In that sense, evolutionary processes are themselves evolutionary products. They vary across the tree in the same ways as organisms and kinds do. Glenn Adelson (forthcoming) has aptly called this attitude to evolutionary processes "Darwinism about Darwinism."

Out of all this, we get a particular picture of the living world. One of that world's contents is a great array of Darwinian populations. These are of many kinds, found at different levels, and include powerful, trivial, and marginal ones. They give rise to the tree, but evolve along with the organisms comprising the tree, being variously suppressed, modified, and augmented, both via endogenous change and through evolution on the part of other populations. The resulting Darwinian picture is, again, partially at odds with old human psychological habits, an old set of responses to biological phenomena, these responses being products of our history of practical dealings with living things—products of our own insertion into Darwinian processes.

CHAPTER 2

NATURAL SELECTION AND ITS REPRESENTATION

2.1. Summaries and Recipes

[*The "classical" tradition; role of abstraction; Weismann, Lewontin, and Ridley; summaries in the form of recipes; two routes to understanding.*]

One way of approaching evolution by natural selection is to try to give an abstract summary of what is essential to the process. This tradition has its roots in Darwin's original discussions, but has become more prominent in recent times. I set out by describing and then continuing this tradition.

Darwin does not begin the *Origin of Species* this way. Instead he begins with empirical phenomena, and works towards his theoretical statements gradually. But by the middle, and especially the end, of the *Origin* he has begun to offer summaries. The most explicit one is given in the final paragraph of the book. Here Darwin sees change by natural selection as the consequence of some simple natural "laws."

These laws, taken in the largest sense, being Growth with Reproduction; Inheritance which is almost implied by reproduction; Variability from the indirect and direct action of the external conditions of life, and from use and disuse; a Ratio of Increase so high as to lead to a Struggle for Life, and as a consequence to Natural Selection, entailing Divergence of Character and the Extinction of less-improved forms. (1859/1964: 489–90)

In summaries like this Darwin is emphasizing two things. One is the fairly abstract nature of the requirements for evolution. There is no mention of the machinery by which reproduction and inheritance occur, for example. The other feature emphasized is a kind of inevitability about the process. If certain preconditions are met, evolutionary change follows inexorably.

Darwin's summaries in the *Origin*, however, are more detailed than modern ones, which often aim for great simplicity. An early summary in this style was given by August Weismann (1909: 50):[1]

---

[1] I am indebted to Lukas Rieppel for bringing this Weismann summary (overleaf) to my attention. Gould (2002: 223) quotes a passage in which Weismann singles out his "extension of the principle of selection to all grades of vital units" as his most important idea.

We may say that the process of selection follows as a logical necessity from the fulfillment of the three preliminary postulates of the theory: variability, heredity, and the struggle for existence, with its enormous ratio of elimination in all species.

Many others have been given since then. Perhaps the formulation that is most often cited is due to Richard Lewontin (1970: 1).

As seen by present-day evolutionists, Darwin's scheme embodies three principles ...:

1. Different individuals in a population have different morphologies, physiologies, and behaviors (phenotypic variation).
2. Different phenotypes have different rates of survival and reproduction in different environments (differential fitness).
3. There is a correlation between parents and offspring in the contribution of each to future generations (fitness is heritable).

These three principles embody the principle of evolution by natural selection. While they hold, a population will undergo evolutionary change.

Here is another formulation that Lewontin gave later, which in some ways provides a better starting point (1985: 76).

A sufficient mechanism for evolution by natural selection is contained in three propositions:

1. There is variation in morphological, physiological, and behavioral traits among members of a species (the principle of variation).
2. The variation is in part heritable, so that individuals resemble their relations more than they resemble unrelated individuals and, in particular, offspring resemble their parents (the principle of heredity).
3. Different variants leave different numbers of offspring either in immediate or remote generations (the principle of differential fitness).

[A]ll three conditions are necessary as well as sufficient conditions for evolution by natural selection. ... Any trait for which the three principles apply may be expected to evolve.

In modern discussions, like Lewontin's, the aim of abstraction is especially prominent. The theorist may start by describing how the evolutionary process works in populations of individual organisms, but then note that the ingredients for evolution might be found in other domains. Entities much smaller than organisms, such as chromosomes or genes, and entities larger than organisms, such as social groups, might satisfy the theory's requirements. Weismann himself saw possibilities of this kind, and used them in his theory of how change works within individual organisms during development (1896). Herbert Spencer (1871) and others saw more extravagant possibilities for the application of Darwinian ideas. The idea of an extension of Darwinism beyond its original domain is almost as old as Darwinism itself.

Lewontin's summaries also emphasize the internal reliability of the process in a particular way. They give a *summary* of the evolutionary process in the form of a *recipe* for change. Variation, heritability, and fitness differences are presented as ingredients. If we mix them together, evolutionary change results. The intent in Darwin and Weismann is similar; the aim is to show that there is a simple machine at the heart of Darwinism. The machine has a kind of causal transparency, and its operation (if properly fuelled) is inevitable. Change results with what Weismann called "logical necessity."

There are also contrasts between the modern formulations and the earlier ones. The "struggle for life," so prominent in Darwin and Weismann, is absent in Lewontin. Vague talk of "inheritance" and "heredity" in the earlier writers is replaced by statistical language. Heredity was always the weak point in Darwin's theorizing, and we see some of this in his summary above. It is not true that inheritance, in the relevant sense, is "almost implied by reproduction." It is entirely possible for there to be reproduction and the reliable reappearance of variation in each generation, without parents resembling their offspring with respect to this variation.

Summaries of this kind do not only appear in theoretical discussions. They are also used to convey some core Darwinian ideas in introductory presentations, and to display the coherence of evolutionary theory in response to attacks from outside. I will round out this sample of summaries with one given in Mark Ridley's textbook *Evolution* (1996: 71–2).

Natural selection is easiest to understand, in the abstract, as a logical argument, leading from premises to conclusion. The argument, in its most general form, requires four conditions:

1. Reproduction. Entities must reproduce to form a new generation.
2. Heredity. The offspring must tend to resemble their parents: roughly speaking, "like must produce like."
3. Variation in individual characters among the members of the population. ...
4. Variation in the *fitness* of organisms according to the state they have for a heritable character. In evolutionary theory, fitness is a technical term, meaning the average number of offspring left by an individual relative to the number of offspring left by an average member of the population. ...

If these conditions are met for any property of a species, natural selection automatically results. If any conditions are not met, natural selection does not occur.

I will refer to formulations of this kind as the "classical" tradition of summary of evolution by natural selection. These summaries tend to have three ingredients: variation, heredity, and differences in reproductive output, though sometimes the ingredients are broken down more finely, as in Ridley. They aim for causal transparency and are often expressed as recipes for change. They describe a mechanism that in the short term does no more than change the distribution of characteristics in a population. These summaries allow that the reliability of inheritance may be weak—there may be only a slight tendency for parents to resemble their offspring—and yet change can occur. And in the modern summaries, though not Darwin's, nothing is said about long-term change, or about whether the change that is predicted will make organisms "better adapted" to their environments.

These summaries are on the right track. However, existing formulations do have problems. It would not be appropriate to apply a critical blowtorch to formulations which are intended to be compact summaries accompanying more extensive discussions, but a close look at them provides a good point of departure and their breakdowns are illuminating. Their usual aim, again, is to describe the Darwinian process by listing a set of ingredients, and noting how they will interact to produce change. But as they stand, the standard summaries do not cover all cases, and do not suffice to predict change. In the next section, and later in the Appendix, I will describe various cases that illustrate this. I also offer a diagnosis of the situation. The standard summaries have problems because they attempt to perform two theoretical tasks at once. Those tasks are (i) describing *all genuine cases* of evolution by natural selection, and (ii) describing a *causally transparent mechanism*. Both jobs are worth doing, but it is hard to do both at once with a single formulation. Existing summaries can be seen, in retrospect, to be sprawled across the two tasks.

2.2. Births, Deaths, and Idealizations

[*Causal and constitutive questions; fitness and discrete generations; interaction of selection and heredity; idealization and understanding.*]

Before looking at the summaries in detail, it is necessary to look at some ambiguities concerning their intended role. First, the usual aim is to give "necessary and sufficient conditions" or just "sufficient conditions" for evolution by natural selection. But this may mean that the task is describing conditions that will *produce* evolution by natural selection (where we know what evolution by natural selection is), or it may mean that the task is giving conditions for something *being a case* of evolution by natural selection. The aim could be to answer a *causal* question about evolution (how does it happen?) or a *constitutive* question (what is it?). Once we make the distinction, we see that the summaries usually try to answer both questions at once. They describe a situation in which a certain kind of change should occur, and the entire process is identified with evolution by natural selection. Summaries are often given in the form of a recipe for change.

There is also a further ambiguity, once we are thinking in terms of recipes. These formulations are usually interpreted as saying that whenever we have variation, heritability, and fitness differences with respect to a particular trait in a population, change in that trait will ensue. But some can also be read as saying that whenever a population features a *general* tendency to show variation, heritability, and fitness differences, then *some* traits will change. The view I will eventually defend, at the end of this chapter, is closer to that second way of thinking, though this is not how the summaries themselves are usually read. I will start by interpreting them in the former, trait-specific way.

There is yet another uncertainty. If someone says they are summarizing *evolution* by natural selection, they clearly aim to describe a process of change. But a summary of *natural selection* might include cases where there is no change, because a population is being held by selection at a particular point—it *would* have changed, if there had been no selection, but it did not. Again, the usual interpretation of the recipes is the former, and sometimes the language makes this clear, but in some ways it would be better to start with the latter.

I will discuss two families of problem cases in this section. The first has to do with reproduction and "fitness." The second has to do with heredity.

Clearly natural selection has something to do with differences in how much individuals reproduce. Many summaries and other discussions specify a way of measuring reproductive differences; an individual's fitness is identified with the *number of offspring* it produces. Sometimes fitness is said to be the "expected" rather than the actual number, and sometimes a relative rather than absolute measure is used. Those distinctions do not matter to this first discussion. Sometimes, as in the Lewontin 1985 summary, more remote descendants are taken into account as well, but here I will initially set aside remote descendants, and first treat the fitness of individuals as measured by the number of offspring produced by that individual, and the fitness of a type as the average number produced by individuals of that type.

This looks like an obvious approach, but counting numbers of offspring is often not sufficient. Starting with the simplest possible example, suppose we have a population of individuals of types **A** and **B** present in equal numbers at some initial time. Every individual divides to produce two offspring of the same type as the parent. Later, all individuals do the same thing again, and again. But **A** individuals cycle through this process twice as fast as **B** individuals do, owing to their more efficient metabolism. So more **A** individuals are produced, and the frequencies of the types change. Although there is change, there are no differences with respect to the *number* of offspring produced by individuals, or produced on average by the different types. The differences concern the rate at which new individuals are produced per unit of *time*.

It might be objected that this is a very unusual case, as the population is growing without constraint. That situation will not last long. But this feature is not essential to the argument. Suppose now that we have a population growing in the same way, living and then dividing, except that as the population gets larger it stretches its resources and grows more slowly. Each individual faces a possibility of death before reproduction, which has the same probability for **A** and **B** types, but which increases with the overall density of the population. There will be evolutionary change in such a situation. But there are no differences between the types with respect to how many offspring an individual has if it survives, or with respect to how likely an individual born at a given time is to survive long enough to reproduce.[2]

Any summary of natural selection that measures fitness differences in terms of the number of offspring produced by individuals will exclude cases like these. This is not a problem for evolutionary theory itself. There are elaborate models of situations of this kind. These are models of "age-structured populations," and in these models survival and reproduction are described in more detail. In a simple case like my first example above, each type (**A** and **B**) is described with an "$l(x)$ schedule," which specifies how likely an individual of that type is to survive to age x, and an "$m(x)$ schedule," which specifies how many offspring an individual of that type will have at age x. From these numbers it is possible to calculate the different "growth rates" of each type, and describe why **A** will increase relative to **B**.[3]

What we see is that many discussions are making a tacit idealization. They treat all cases of natural selection as if they occurred in situations in which generations are *non-overlapping* and *synchronized* across the population. This is often called a "discrete generation" model of evolution. These are the simplest cases to analyze. And some organisms do have non-overlapping generations synchronized across the population. These include annual plants such as basil, many insects, and some others. But most organisms do not reproduce like this; humans obviously do not. Evolution is, of course, a process that takes place in time. In some cases, the role played by time is made so simple by the life-cycle of the organisms that

[2] Here is a bare-bones model of such a case. The expected fitness for each type is $W = 2(k - N)/k$, where k is a constant and N (always less than k) is the total size of the population at an individual's birth. This formula describes both the **A** and **B** types, and N is the same for **A** and **B** individuals born at the same time. But given the difference in speed of reproduction, the frequencies of the types will change, at least until N reaches an equilibrium value where it is equal to $k/2$, at which point evolutionary change stalls because the reproductive events have no effect on overall numbers. If the population is growing from below the point where $N = k/2$, the frequency of **A** increases until the "stall." If the population is shrinking from above that point, the frequency of **A** decreases.

[3] To find the growth rate, λ, of a type (once the population has reached a stable age distribution) we solve the following equation for the type: $1 = \sum_x \lambda^{-x} l(x) m(x)$. Suppose, for example, that time is measured in days, and the **A** types always live for one day and divide into two at the end of that day, while the **B** types always live for two days and then divide. Then the growth rate per day for **A** is 2 and that for **B** is $\sqrt{2}$. These numbers can be used to predict the rate at which the frequency of **A** will increase relative to **B** in the total population (Crow 1986: ch. 6).

time need not be explicitly mentioned. But a summary that ignores the role of time can only be seen as describing an *imagined simpler relative* of the process of evolution in the majority of cases.

This argument is simplest when we think of the fitness of an individual as its number of immediate offspring. Things are less clear when "remote" descendants are included, as in Lewontin's 1985 summary. Remote descendants can serve as a kind of proxy for time. In my cases above, for instance, the frequency of the A type increased relative to B due to its faster rate of reproduction. That means that across any specified time interval, the initially present individuals of the A type will finish with more remote descendants (great-great- ... -grand-offspring) than initial individuals of the B type, because the A individuals complete more generations in the available time. In my second example I supposed that the likelihood of early death increases gradually as time passes. Then although any pair of A and B individuals born at the same time have the same expected number of offspring, the B individual's offspring will be born during tougher times than the A individual's offspring, so the B individual should have fewer grand-offspring (and great-grand-offspring) than the A individual.

The mention of grand-offspring in discussions of fitness is usually intended to pick up some special cases in which the advantage associated with a trait does not show up in the numbers of immediate offspring, but does in grand-offspring. An example is Fisher's (1930) explanation of why sex ratios are usually roughly 1:1. Individuals producing the rarer sex may not tend to have more offspring, but will tend to have more grand-offspring, because their rarer-sex offspring will be in demand for matings. In the cases above, in contrast, the remote descendants are serving as longer-term indicators of features that A-type and B-type individuals manifest during their own lives, their ability to reproduce at faster or slower rates.

The next move might seem obvious. As most populations are age-structured, we should apparently borrow a measure of fitness from this part of evolutionary theory. We can say that evolutionary change is driven by fitness differences, but with a different understanding of fitness. However, the models of age-structured populations make their *own* simplifying assumptions. Things that are easy to handle in the discrete generations case—like sexual reproduction—become difficult in the age-structured models. The "rates of increase" described above are not always useable as a measure of fitness. The simple problem cases in this section are the tip of a rather strange-shaped iceberg. According to Brian Charlesworth, who has developed and surveyed these models extensively, there is *no* single parameter that can be seen as the "fitness" of a type in a way that predicts change in all cases (1994: 136).[4]

---

[4] I should quote the Charlesworth claim in full, so it is clear what he is and is not saying: in an age-structured sexual population "no single parameter can be regarded as the fitness of a

What does this show? The situation looks very different from the practical and the foundational points of view. From the practical viewpoint, the consensus seems to be that a good approximate description of a case can usually be achieved either by assuming a discrete generations model or, in some cases, an age-structured model, with each making different simplifying assumptions (Crow 1986: 175). A modeler can pick and choose between tools, depending upon convenience and the details of the case at hand. The ability to make such choices is part of the "craft knowledge" associated with this part of science. If our aim is to say what evolution by natural selection *is*, however, and what role "fitness" has in the process, then the idealizations that are habitually made pose a problem.

I now move to a second family of problems. These concern the role of heredity. Looking back at the previous section, we see that Darwin and Weismann talked in an informal way about "heredity" and "inheritance," while Lewontin sharpened things up by using a statistical concept of "heritability." I will use the term "heredity" in a general way to refer to many phenomena involving parent–offspring similarity and the inheritance of traits, and "heritability" to refer to a family of statistical measures (discussed in the Appendix). These describe the extent to which, in a population at a time, the state of a parent is predictive of the state of its offspring with respect to a particular trait.

It is important to see what is achieved by the introduction of heritability. If we want a recipe, a predictive formula, then we need *some* exact measure of the parent-offspring relationship, not just vague talk about heredity. We also want this measure to be neutral about the mechanisms of inheritance. What seems needed is a statistical concept. And the concept used must enable us to deal with "continuously varying" traits, such as height. The simplest way to think about evolution is to assume a population divided into distinct types, like the A and B types above. But that is not always possible. Suppose we have a population in which there is variation in height, but no two individuals are the same height, and the gaps between individuals are evenly distributed. Now suppose that the tall individuals reproduce more than the short ones, and the tall ones also tend to have offspring that are taller than average. That is, there is some parent–offspring correlation with respect to height. Then height in the population can evolve. Evolution by natural selection does not need there to be distinct "types" in a population; evolution is possible if everyone, in both parent and offspring generations, is unique with respect to the evolving trait.

This makes heritability look like the right concept to use. Heritability comes in degrees. There may be only a tiny tendency for parents to resemble offspring.

genotype with arbitrary selection intensities" (1994: 136). For more of the iceberg see Beatty and Finsen (1989) and Ariew and Lewontin (2004).

If so, then even if there are huge differences in fitness, there will only be a small amount of change. Heritability determines how a population will "respond" to fitness differences.

The problem is that there is a variety of cases where there can be variation, heritability, and fitness differences but no evolutionary change. In discussing them I will assume discrete generations and make many other simplifying assumptions; the problem does not come from the idealizations discussed earlier. The problem comes from the fact that heritability is *so* abstract a concept; it throws so much information away. As a result, there can be a kind of "canceling" of the roles of fitness differences and the inheritance system, in a way that results in no net change.

Here I will give a very simple example which shows one variety of the phenomenon in a vivid, though unrealistic, form. Suppose a population contains short (height = 1), intermediate (height = 2), and tall (height = 3) individuals. Intermediate individuals are fitter than either extreme. An illustration is given in Figure 2.1. Heritability is high, but the parental and offspring generations are identical. This is because selection favoring the intermediates is exactly compensated by a dispersing tendency in inheritance, found just in those intermediate individuals.

This case is a toy one, but the principle it illustrates is real. The idea motivating use of heritability is that if there is *some* tendency for parents to resemble offspring, by whatever mechanism, then a population will respond to fitness differences by changing across generations. But the specific patterns of heredity underneath the general fact of "parent–offspring similarity" can be such as to nullify the effect of fitness differences, rather than transmit them to the next generation.

There is no deep puzzle here. In the case in Figure 2.1, the pattern of heredity was *already* tending to produce change, and the fitness differences pushed in the other direction. Not all the problem cases have this feature, as we will see

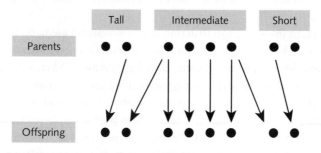

Figure 2.1: Stabilizing selection in an asexual population.[5]

---

[5] You might find an optical illusion in the figure, in which the lower "flanking" individuals seem pulled in towards the central group.

in the Appendix, but there is usually something related going on. A look at the details removes any appearance of paradox. But from a foundational point of view, we do have something to contend with. The aim was to see variation in character, heritability, and fitness differences as ingredients that, when combined, necessarily give rise to change; these are what makes evolution "run." When I say "necessarily" just above, this does not mean it was being claimed that *no* other factors could intervene to prevent change. But the requirement of heritability was supposed to summarize what is needed with respect to the pattern of inheritance. That view is now seen to be too simple.

Looking at both sets of problems discussed in this section, here is a diagnosis of the situation. The standard summaries are often the products of trade-offs between two different approaches to scientific understanding. To achieve understanding of some class of phenomena we might do either, or both, of two things. One option is to try to give a literally true description of the important features of all cases of the phenomenon. The other option is to give a detailed description of one class of cases, usually a relatively simple class, and use these as the basis for an understanding of the others. The exact description we give for the simple cases is *known not to apply*, in a direct and literal way, to the complex ones. Understanding is achieved via similarity relations between the simple cases we have explicitly analyzed and the more complex ones. The class of cases illuminated by their similarities to the well-understood ones may be vague and open-ended.

Within this second strategy, the cases described directly might either be a simple subset of the empirical cases, or a set of fictional cases arrived at by imagining away some factors. That imaginative act is what I referred to as "idealization" earlier in this section. Either way, the simple cases function as *models* for others, so this can be called a model-based strategy of theorizing (Giere 1988, Godfrey-Smith 2006, Weisberg 2007). A person can employ both of these strategies at once, but each goal exerts a different kind of pull on any single description.

The problem involving age-structure is an interesting case here. Some actual populations do have discrete generations. Others, like ours, have complex age-structure and breed at different times over long periods. When someone treats a discrete-generations model as saying something about human evolution (for example, when they use it to explain the retention of different traits in a stable equilibrium), they are treating the actual human population as relevantly similar to an imagined population that has discrete generations. And for practical purposes, this often works well. It is like describing how objects fall to earth using a model that treats air resistance as having no effect.

The various summaries and their relations to problem cases will be discussed again in the Appendix. The formulas bump into different combinations of difficulties, depending on the wording. The overall picture, though, can be summarized like this. Many summaries of the Darwinian process aim to do two things at once. One is to say what evolution by natural selection *is*. The other is to display a kind of causal transparency in the Darwinian machine; thus the long history of claims of the form: "*if* you have these few simple ingredients, then change must result." The classical summaries are my starting point in this book; they are set up along roughly the right lines. But the formulations that have been given tend to be the products of trade-offs described above. As the summaries get sharper and function better as recipes, they start to omit cases. As they become more inclusive, they break down as recipes.

2.3. Fitness, Drift, and Causation

[*Selection and causation; trait-by-trait versus whole-organism analysis; fitness-related properties; organization of evolutionary theory; happenstance change as a default.*]

A central part of the modern treatment of evolution is the distinction between change due to selection, and change due to "drift"—change due to differences in survival and reproduction that merely reflect the operation of chance. The relationship between selection and drift is the topic of formidably complex mathematical models and a philosophical literature that grows without limit. In the next chapter I will offer a new treatment of the distinction, using a framework developed there for other reasons. But it is necessary to say something about it now, as the topic generates problems with the attempt to give a summary.

The simplest way to see the problem is to look at the Lewontin formulations quoted at the beginning of this chapter. Lewontin (1985) says there is change by natural selection if "different variants leave different numbers of offspring either in immediate or remote generations". But, it seems, the mere fact that some variant (the **A** type, perhaps) leaves more offspring than another (**B**) does not mean that natural selection was operating. For all we know, the difference between them could be due to chance. Many of the **B** types could have been struck down by various kinds of accident (lightning strikes, earthquakes, localized epidemics ...) before reproducing. The fact that the **A** individuals did better under the threat of lightning and epidemic *may* reflect some natural advantage they have, but it also may not. It might be simply a matter of luck.

So the classical summaries seem incomplete unless they make some distinction between reproductive differences due to the advantages bestowed by phenotypic characteristics, and reproductive differences due merely to accident. But it is not clear how to do this, for several reasons. One is the fact that the concept of causation—which seems central here—is philosophically vexed and also treated with caution by many scientists. (The caution is visible in Ridley's summary, which requires fitness differences "according to" phenotype—one of those phrasings that balances precariously between causal and statistical language.) A second is a suspicion which will be discussed in Chapter 3. When we talk of a category of "random" or "accidental" change, don't we really just mean that the causes for those reproductive differences are complex, fine-grained, and currently unknown to us? Is "drift" a name for a category of change in nature, or just a label we use to reflect the limits of our knowledge?

The discussion in this section will be a preliminary one. The first thing to do is distinguish between two ways of describing an evolutionary process. One can describe evolution in a trait-by-trait way, or by thinking of whole organisms as units and not breaking them into "traits." Start by thinking in a trait-by-trait way. For any particular trait that changes in a population, the question can be asked: was the change due to selection acting on that trait, or did it change as a byproduct of some other process? The question makes sense because any specific trait might have no casual impact on a process of evolution that affects it. For example, there might be selection for trait T_1, and a correlation due to the developmental or genetic system between T_1 and T_2, leading to change also in T_2. T_2 itself may have no causal role in this process; it might be the color of some internal structure that is never normally seen, while T_1 is some functional property of the same structure. Sober (1984) described this distinction by saying that there can be selection *of* something without selection *for* it. This is a case in which change in the frequency of T_2 is in a sense "accidental," though not random. These are not cases of drift, but of "correlated response." There are also other ways for the frequency of a trait to change in a population without selection acting on it. Mutation might be introducing new cases of the trait faster than it removes them, for example. Or migrants bearing the trait might be entering the population.

Is it possible for the frequency of a trait to change *purely* by accident, because the bearers of the trait "happen" to have more offspring? Remaining with the example of the color of an internal organ which is never normally seen, suppose there is a population in which the individuals all have either one color (*red*) or another (*blue*) in this part of their anatomy, and the color difference has no functional consequences and is not correlated with any functionally important trait. The trait has a simple genetic basis, however. Now suppose the *red* individuals have more offspring than the *blue* ones, leading to an increase in the frequency of *red*. Assuming a deterministic world (or near-enough to one), the difference between these reproductive rates will have *some* causal explanation. This *blue* individual was struck by lightning; that one was hit by an errant bus. In these events, the *red/blue* phenotype is disconnected from the reasons for the change, but there is *something* about each *red* individual, and each *blue* individual, that is responsible for its doing well or badly. Maybe it is a matter of properties that seem fortuitous. Maybe it is a matter of being in the right place at the right time. But if we have a sufficiently broad conception of the total set of "characteristics" of an individual, then its reproductive output is a consequence of its characteristics.

So if a summary is set up in a trait-specific way—a summary of what is involved in change via selection *for F*—then we have to deal with the fact that any trait can change through differential reproduction and heritability without itself being selected for. Consequently, a summary of this kind must include a requirement that gives F a suitable causal role. This extra requirement is not hard to express. Change via selection for F requires not just variation in F-ness, heredity, and different rates of reproduction. It also requires that the variation with respect to F be partly causally responsible for the differences in reproductive output (Hodge 1987, Millstein 2002). However, it is not so clear that something like this is needed if a summary is not aimed at saying what is involved in change via selection for F. I leave the topic there, though, as it will be taken up again in the next chapter.

In the rest of this section I will discuss some more general issues related to fitness, causation, and the organization of evolutionary theory.

There is a huge literature on the nature of fitness. The word will be used constantly in the rest of the book, so I will say a bit about how it is being understood. Much of the literature on fitness has been an attempt to say what the fitness of an organism or type of organism *really is*. Darwinism is seen as committed to the idea that fitness is some unified and definite property, elusive to analysis, but the real driver of evolutionary change. That is not my approach. Some reasons for having a different view have been introduced already, but to make the issue clear I will discuss here, as a foil, the most prominent approach to analyzing fitness in recent years, the "propensity view" (Brandon 1978, Mills and Beatty 1979, Sober 1984).

The propensity view holds that fitness is the "expected" number of offspring had by an individual, or by individuals of a given type. This is contrasted with the "realized" fitness, the number actually produced. A controversial concept of physical probability is used in the calculation of the "expected" value, and that has been a focus of dispute. The hope has been that *if* such an appeal to probability is defensible, then the propensity view can give an account of the properties of organisms that drive evolution by natural selection. Fitness measures the level of reproductive success that an organism's biological properties would *tend* to bring about, given the environment and the present composition of the population. This makes it possible to see fitnesses as reflecting the overall level of adaptation that organisms have to their circumstances, and it makes sense of the idea that change can, through accident, go in the opposite direction from that predicted by fitness differences.

As we saw, however, if the aim is to isolate the properties of organisms that drive evolution, then tracking *numbers* of offspring (whether "expected" or "realized") is often not enough. In many situations, for example, the timing of reproduction matters as well as the numbers. The list of odd effects in this area is now a long one. Usually reproducing earlier is "better" than reproducing later ... but not if the population is shrinking (see Charlesworth and Giesel 1972, Ariew and Lewontin 2004, and footnote 2 above). Even in the simple case of discrete generations, a type with a lower average fitness can beat a type with higher, if the lower-average type also has lower variance (Dempster 1955, Gillespie 1972, Frank and Slatkin 1990). So even if the controversial assignments of probability used by the propensity view are available, the resulting propensities have a restricted importance.

One response to this situation is to try to come up with an even more refined measure of what the real fitness of an organism is, perhaps a more complicated "propensity," but I think this is unnecessary. It is better to say, instead, that there is a family of fitness-like or fitness-related properties, all involving reproductive output in some sense or other. Different ones are relevant in different circumstances, and this shows up in the diverse fitness-related properties seen in different formal models. Several people have been moving towards versions of this idea.[6] Here is how I think it is best expressed. There is a totality of facts about survival and reproduction, for any population over a time interval, which are responsible for how it evolves. This totality includes facts about the distribution of ages and population growth, as well as about when and how prolifically a given kind of individual is liable to reproduce. Facts about fitness, traditionally, are seen as a sort of "compression" of this totality, one which includes the information necessary to predict change. But in different cases, different elements of this total body of information become necessary in working out what will happen, and often the needed information far outruns anything that might be regarded as a numerical measure of overall adaptedness. So to say that there is a "family of fitness-related properties" is not to move towards a soft-minded conventionalism on the question of how evolution runs. It is, instead, to recognize that "fitness"-talk involves a compression of a full specification of causal factors, and such compressions have limitations that manifest differently across different cases.

This attitude towards fitness can be connected to a general picture of the organization of evolutionary theory. Speaking very broadly, we can think of evolutionary theory as containing a collection of abstract models of basic processes,

---

[6] These include Beatty and Finsen (1989), Day and Otto (2001, a mild version), and Krimbas (2004, a more radical one). See also the Charlesworth quote in footnote 4. Some would say that survival-related properties must be included as they count in their own right, not just as they affect reproduction. This will be discussed in Sections 2.6 and 5.4.

plus a mass of mid-level theory connecting these models to the features of actual organisms and the history of life on earth. The collection of models forms a patchwork, adjoining and overlapping, each dealing with the complexity of evolution in different ways (Potochnick 2007). Usually, to represent one set of factors in a comprehensible way it is necessary to background others. This "backgrounding" is partly a matter of ignoring things, but partly of idealizing—imagining features away. The patchwork of models, and accompanying commentaries, express our dynamic understanding of evolutionary processes, our knowledge of what will give rise to what. When trying to give a foundational description of evolution, it is common to seize on some particularly useful-looking part of the patchwork, and treat it as the basis for a description of all cases. Returning to the main themes of this chapter: a good *recipe* for evolution will do exactly that; it will seize on some part of the patchwork which shows the operation of a Darwinian machine in an especially transparent way. It will show how the wheels can turn, how collections of insignificant-looking events can have large-scale consequences. But this does not mean that the model, once verbalized, can function as a summary of all cases.

Here I have emphasized the role of fitness, but there are other illustrations as well. A model may assume stasis as a default, treating selection as a force which intrudes to produce change. That is clear and convenient, but there is no reason to see this as reflecting how things are in nature. If we ask what the "default" assumption is for a population in nature, the answer is surely that such a population will ordinarily be undergoing a sort of happenstance change. Some individuals will live longer and reproduce more than others. Some traits will become, in the short term at least, more common. We do not have to introduce a theoretical concept of natural selection to note this; it is an expected feature of the noisy natural world and not something that Darwin should be seen as discovering (Kitcher 1985). What we would expect is a perpetual rustle and murmur of low-level change. Darwinian theory, and some rivals to it, enter as collections of claims about whether this happenstance change will amount to anything significant, and why. An essentialist "natural state" model will hold that a population may change a bit but will tend back towards its prior and proper condition. Darwinism, in contrast, holds that given certain tendencies in these low-level events, we *can* expect the population to go somewhere, and often to produce something new.

2.4. The Replicator Framework

[*Formulations in Dawkins, Hull, and others; replication not required; types and quantitative traits; sex.*]

All the ideas discussed so far in this chapter belong to what I am calling the "classical" tradition. For some time there has been an alternative, the *replicator* framework, favored by Dawkins, Hull, Dennett, Haig, and others.[7] The relationship between the two approaches has always been unclear. Sometimes they are seen as roughly equivalent, with the replicator view giving us a shorter and more colorful description. There are other ways in which the two can be seen as compatible, which will be discussed below. Often, though, the replicator view is presented as a complete analysis of how natural selection works, superseding other views, including (presumably) those discussed above. Initially, that is how the replicator view will be treated here, and when understood in those terms it will be criticized fairly vigorously. But some thinking behind the replicator approach does have a useful role in a more restricted context. When I put my own view together later in this book, some themes from this literature will reappear.

In its original forms, the replicator view holds that all cases of evolution by natural selection involve the operation of replicators. This is a role that *some entity must always fill*, as a matter of principle, though different entities can play this role on different occasions. In most discussions there is a second and equally important role that something must play, that of an *interactor* (for Hull) or *vehicle* (for Dawkins), which will not be discussed until Chapter 6.

So what is a replicator? Dawkins takes the idea of *copying* for granted, saying in one of his fuller definitions: "We may define a *replicator* as any entity in the universe which interacts with its world, including other replicators, in such a way that copies of itself are made" (1978: 132). Hull says a replicator is an "entity that passes on its structure directly in replication" (1980: 318). Later, Maynard Smith and Szathmáry gave a quite different definition: a replicator is anything that "can arise only if there is a preexisting structure of the same kind in the vicinity" (1995: 41). They add that, within this broad category, a *hereditary* replicator is one that can exist in several different forms, where these differences are passed on in replication. My discussion of replicators in this chapter will mostly focus on the views seen in Dawkins and Hull. The definition given by Maynard Smith and Szathmáry, which I see as different in spirit, will be discussed in a later chapter.

The classical approach can be seen as taking individual organisms as its starting point, and introducing the possibility that other entities could play similar evolutionary roles. They might be smaller than organisms, like chromosomes, or larger things like colonies or groups. Individual organisms are the *original* domain covered by the classical concept, but they might not be the *only*, or the most important, domain. The replicator view, in contrast, takes genes—alleles—as its starting point. Dawkins in 1976 argued that taking a

---

[7] See Dawkins (1976, 1982a), Hull (1980, 1988), Hull et al. (2001), Lloyd (1988, 2001), Dennett (1995), Maynard Smith and Szathmáry (1995), Sterelny et al. (1996), Haig (1997). Maynard Smith and Gould are notable for endorsing *both* approaches (Gould 2002: 609, 615; Maynard Smith 1988). This is possible, as we will see, though as far as I know neither author discussed the relation between the two in detail.

"gene's eye view" of evolution helps us with many theoretical problems, especially involving altruism. Once we see how these analyses work, he argued, we see the possibility of a new foundational description of evolution, and also see that different things could, in principle, play the role that genes play in the familiar cases. We may decide that genes are not the only possible replicators, as it is the replicator *role* that matters.

I will give two sets of arguments against the replicator analysis. The first argues that the replicator analysis does not cover all cases (Avital and Jablonka 2000, Godfrey-Smith 2000). The second criticizes the use of "agential" concepts in many (though not all) versions of the replicator view.

The first argument is simple. Although the details are sometimes unclear, a replicator is supposed to be something that is accurately or faithfully copied, and can form lineages of such copies over long stretches of evolutionary time. Replicators are not supposed to be copied with 100 percent accuracy, as that would prevent the appearance of new variants, but the replication process is supposed to be a high-fidelity one. In Section 2.2, I introduced the case of a population in which there is variation in height, a reproductive advantage associated with tallness, and moderate but imperfect heritability of height. Initially, suppose that reproduction is asexual, so each individual has only one parent. Taller than average individuals produce taller than average individuals, but the variation in height is fairly evenly spaced. No two individuals are the *same* height, either within or across generations. So organisms in this situation do not "replicate" themselves; they do not "pass on" their structure or type. But evolution can certainly occur. When the taller ones reproduce more than the shorter ones, and the taller ones tend to have taller offspring, the population will change its distribution of heights over time.

Replicator analyses usually do not suppose that whole organisms are replicators. The claim, rather, is that the inherited differences between individuals should be due to replicators *somewhere* in the system. But now suppose, in the case of some continuously varying trait like height, that the mechanism of inheritance is not one that involves "copying" at the lower levels either. Everywhere we look, there are degrees of similarity but no variation "faithfully transmitted." We do not have to suppose that *nothing* is being copied in the organisms in question, just that the mechanisms responsible for passing on these differences between individuals do not involve differences in what is copied. There are specific possibilities that can be imagined here, including some cultural inheritance processes (Chapter 8), and cell-level inheritance systems.[8] But the important point is that it does not *matter*

---

[8] Known real-world examples which resist description in terms of the "copying" of a structure in this way are limited, but include some "structural inheritance" systems in single-celled organisms and inherited changes to gene expression levels and chromatin structure in plants and

what is imagined at this lower level, because all we need is that reproduction lead to parent–offspring similarity at the level at which evolution is to occur. It does not matter what particular mechanism underlies this pattern of similarity, so long as the pattern is present.

The relevant sense of "similarity" will be discussed further in the Appendix. Roughly, what is needed is that the state of parents correlate with that of their offspring; parent and offspring must be predictive of each other to some extent, or more similar than unrelated individuals. This parent–offspring similarity is not always sufficient to produce change—there is the possibility of the fitness differences and the pattern of inheritance canceling each other, as discussed above. But weak parent–offspring similarity when combined with selection is often enough.

One response that might be made at this point is that to get *significant* or *sustained* evolution, there must be replicators, even if some evolution is possible without them. This idea will be discussed in the next chapter, but the initial point to note is that it is changing the target of the replicator analysis. The original claim was that *any* process of evolution by natural selection, as a matter of principle, has to have something in the replicator role, not that the "important" ones do. In this chapter I am just concerned with the "*any* ..." question.

There is also a more subtle feature of the replicator analysis' insistence on high-fidelity copying that is significant. A feature of both the definitions that are usually given, and also of the way that the concept is actually used, is that each replicator is of some particular *type*, and this type is passed on reliably through copying. Different varieties of replicator compete with each other, and evolutionary change is understood as change in the frequencies or numbers of copies of the various types. This is the simplest way of thinking about evolution, and it is used often in this chapter. ("There is the A type and B type, and the A types become more common because ... .") One way to think about my example using variation in height is to note that this "division of the population into types," as well as the faithful transmission of type, is not necessary. One individual may be more or less similar to another, with respect to height, without there being a type that is passed on or not passed on. However, types are often (not always) presented as integral to the replicator view. Replicators are supposed to form "lineages of copies"; type-hood is transmitted over time in the form of causally connected material tokens. I, in contrast, will make much of the simple phenomenon described above: the possibility of evolution in a situation where *everything is unique*, and is treated as such.

other organisms (Grimes 1982, Hollick et al. 1997, Moliner et al. 2006). For a general review of "epigenetic" inheritance systems see Jablonka and Lamb (1995). A simpler example is provided by cultural inheritance in which ideas or behaviors are not transmitted intact but, through influence on an observer, as similar levels of some continuous variable (Chapter 8).

This view is a kind of "evolutionary nominalism." Stated explicitly: the grouping of individuals into types is in no way essential to Darwinian explanation. Such groupings are convenient tools. But one always has the choice of using finer or coarser groupings, ignoring fewer or more differences between individuals. As categories become finer, they may be occupied by only one individual each. This does not make evolutionary description collapse; it is possible to describe similarities in a population without grouping it into types at all, by using orderings and metrics (as we do with height).[9]

At this point some advocates of the replicator view might insist that there was nothing beside convenience behind the formulation in terms of high-fidelity copying and faithful transmission of type. In a situation in which there are only degrees of similarity, there can still be replicators. Perhaps an entity can be a replicator if it is causally responsible for things that are similar, to some relevant degree, to itself.[10]

This is a start—it is a move back towards the classical view. But the way the replicator framework is set up makes the required move awkward. The replicator view is normally expressed in a way that assumes that replication is asexual. The whole language of "passing on structure" is based on a picture in which a replicate has one parent, precursor, or template. But many ways in which new entities are produced in an evolutionary context are sexual; they involve the action of two parents or precursors. That is true not only with multicellular organisms like us. Even viruses can reproduce sexually. If two particles infect a single cell their genetic material can be mixed in the production of new viruses (Foissart et al. 2005). If something is the offspring of two parents, and the parents

[9] The idea that evolutionary change is necessarily a matter of types is not confined to replicator views. "Selection theory is about geno*types* not geno*tokens*" (Sober and Lewontin 1982: 172). The thought underlying some of these claims may be that without a grouping into types, evolutionary description collapses. ("Natural selection is the process by which replicators change their frequency in the population relative to their alleles. If the replicator under consideration is so large that it is probably unique, it cannot be said to have a 'frequency' to change." Dawkins 1982a: 88). This is one reason that formal models that do not require a grouping of the population into types, such as the Price equation (A.1 below) are important. Mayr saw a view of evolution as "the preservation of superior types and the rejection of inferior ones" as part of the typological mode of thinking he opposed (1976: 159; see also Nanay, forthcoming).

From a formal point of view, a partition of a population into types or classes is a special case of description of the population in terms of relations between its members ("equivalence relations," which are reflexive, symmetic, and transitive, divide a population up into non-overlapping classes). The theme, seen here, that we tend to think in terms of types even when the underlying network of relations resists this, is also found in other evolutionary contexts. These include problems with groups as higher-level units of selection (6.2 below, and Godfrey-Smith 2008), with populations as units within species (Gannett 2003), and with species themselves (Franklin 2007).

[10] This may be intended in Hull (1981), and an analysis offered in Godfrey-Smith (2000) has this structure. See also Nanay (2002).

differ significantly, then it will not end up *very* similar to either parent, but may still be *more* similar to both parents than to other individuals in the population. Once again, different rates of reproduction can give rise to evolutionary change.

From this point of view, we can see the replicator analysis as picking out a special case of what is covered (or supposed to be covered) by the classical view. The classical view talks of populations in which there is reproduction and heredity. The reliability of heredity can be low, and there can be more than one parent for each individual. The replicator view picks out a subset of these phenomena: Darwinian populations with high-fidelity heredity and asexual reproduction. This is, both in the actual world and in abstract principle, an important subset of the larger category, but it *is* just a subset. Replicators that vary in fitness are sufficient for evolution by natural selection (give or take some special cases), but they are not necessary.

2.5. Agents and Interests

[*Agential views of evolution; fallacies involving persistence; replicators as a model case.*]

I now move to a different set of objections to the replicator approach. But the features criticized in this section are not found in all versions of the view. They are very prominent in Dawkins, Dennett, and Haig, but not at all in Hull, Maynard Smith, Szathmáry, Sterelny, and some others.[11]

The replicator approach, in many versions, is designed to mesh with an "agential" way of looking at evolution, a perspective in which we see the entities in an evolutionary process as pursuing goals, having interests, and using strategies. Both the proponents and critics of this way of talking agree that it is derived from the way we normally talk about entire intelligent agents, and is *some* sort of metaphorical or analogical extension of that framework. There is variation within the proponents of this outlook regarding how seriously the language is to be taken—whether it is an obvious metaphor that is designed to be worn lightly and quickly discarded when trouble looms, or a form of description that has metaphorical *roots*, but functions to pick out an important natural kind, involving a subtle form of "directedness" seen both in the case of whole agents and genes. I won't worry about this (very interesting) variation here, and will assume that the agential language is intended to be purely heuristic in role. My

---

[11] Hull et al. (2001: 514) explicitly distance their version of the replicator view from any appeal to benefits and goals. In contrast, Lloyd (2001) has defended a version which seeks to analyze and domesticate talk of the "beneficiaries" of evolutionary processes. I think these concepts will always remain unhelpfully feral.

argument, instead, will be that this way of approaching evolution structures people's thinking more than they realize. When we think about a domain in this way, we let loose a particular set of psychological habits, deploy a particular set of tools. The results, I will argue, are often not good ones.

Richard Dawkins famously argued that various things cannot be "units of selection" because they are too temporary (1976: 34).

In sexually reproducing species, the organism is too large and too temporary a genetic unit to qualify as a significant unit of natural selection. The group of individuals is an even larger unit. Genetically speaking, individuals and groups are like clouds in the sky or dust-storms in the desert. They are temporary aggregations or federations. They are not stable through evolutionary time ...

For Dawkins, we tell evolutionary stories in terms of the activities of persisting things. Evolution by natural selection is a process involving entities that change their frequencies over substantial periods of time. So various entities that we might have seen as "units" of selection are in fact disqualified, because they are too ephemeral. Bluntly put, "You cannot get evolution by selecting between entities when there is only one copy of each entity!" (Dawkins 1976: 34).

But that claim is not true. Let's return to the simple cases involving variation in height, assuming either sexual or asexual reproduction. No two individuals are of the same height, within or across generations, and the intervals are evenly spaced so there are not any "clumps." Everyone is unique; there is "one copy" of each entity, but evolution by natural selection is entirely possible.

What is true is that *if* we want to describe evolution using an agential framework, where a process extended over long periods is described in terms of the pursuit of goals by some entity, then that entity must persist, at least in the form of copies. Otherwise it cannot realize or fail to realize its goals. So *within* the framework that collapses populational processes to the activities of agents, Dawkins is expressing a real constraint. But that is an argument against the agential framework—an argument that it cannot be applied to all cases. It is not an argument that evolution cannot occur unless long-term persisting entities are present.

So it is not just overtly teleological features that the agential framework imputes to evolution—features which we can quickly say are only being metaphorically attributed. Subtle structural features are being imposed as well. The agential framework treats the evolutionary process in terms of the persistence of special entities through superficial change, rather than in terms of the successive creation of *new* entities, with similarities and other dissimilarities to earlier entities.

I conjecture that there may be an interesting psychological phenomenon at work here. In this book I try to defend and extend a "populational" point of view on evolution, as outlined in Chapter 1. When thinking about complex systems that exhibit apparent phenomena of adaptation and design, it seems psychologically natural to us to use a different set of concepts, to organize our thinking in terms of persisting agent-like entities of some sort or another. One attraction of the replicator framework is its removal of a sea of transients from the evolutionary center-stage, in favor of a set of hidden, coherent, and persisting things that can be the locus of attributions of agency. These attributions impart a kind of order and comprehensibility on the evolutionary process.

The discovery of lineages of faithfully copied stretches of DNA is empirically very important, of course, and in many contexts *does* belong in center-stage. But some thought-experiments are instructive here. Suppose it had turned out that DNA does not physically persist throughout the life-cycle of organisms like us. Suppose DNA was present in the zygote, was used to form an initial stock of protein molecules needed for development, and was then broken down for use as food. The bulk of an organism's life was carried on in the absence of DNA, via a complex catalytic network of protein action (including, somehow, the manufacture of more protein and the coordination of its activities). When the time comes to make sex cells, DNA is "reverse-translated" from a key set of proteins, and is used to initiate the next generation. DNA would still be the vehicle of inheritance, but would come and go like the rest of the organism.

"Reverse-translation" of protein to DNA is widely believed to be impossible, and I am not saying that such a form of life would be feasible given actual-world biochemistry. But scenarios like this are useful psychological tools for pushing against the tendency to think that it is *essential* to an evolutionary process that something *must persist* rather than being periodically reconstructed (Oyama 1985), that some enduring agency must continue underneath the procession of unique things, underneath the transients and ephemera.

That concludes my initial discussion of the replicator view. A few final comments should be made about versions of the view that would not be subject to the criticisms above. First, a person might say that although the replicator analysis does not cover every case, it provides us with a good model. Earlier I allowed that models of this sort can yield genuine understanding, and argued that there are hidden idealizations in the classical summaries. Might the same not be said in defense of the replicator view? In reply I say yes, thinking about replicators can give us a good model case. Some formal treatments of evolution use a mathematical structure called the "replicator dynamics" to illustrate basic processes without claiming that the analysis applies literally to all (Nowak 2006; see Section A.1 below). Replicators can provide a useful model, but a key feature of models is that we can employ several of them. Another very instructive model case is the phenomenon of evolution in a quantitative character, such as height. These remind us that any mechanism of heredity will suffice and evolution can occur in a population of unrepeated things.

Lastly, I mentioned above the fact that a person might claim that although evolution could *occur* without replicators, these cases will be weak and insignificant. Replicators are needed for evolution to produce anything important. That is the right next question to ask about replicators, and will be one topic of the next chapter.

2.6. The Minimal Concept Summarized

[*Darwinian populations in the minimal sense; an associated category of change; Darwinian individuals; selection without reproduction.*]

Here is a way of putting together the ideas developed so far. A *Darwinian population in the minimal sense* is a collection of causally connected individual things in which there is variation in character, which leads to differences in reproductive output (differences in how much or how quickly individuals reproduce), and which is inherited to some extent. Inheritance is understood as similarity between parent and offspring, due to the causal role of the parents.

What is being specified here is not a recipe for change, but a "set-up," a way in which things may be arranged. The features used to specify the set-up are distinguished, however, by the fact that they are central to the occurrence of particular kinds of change. So while the elements used are drawn from the classical summaries, rather than being used in a recipe they appear in a kind of distillation. Any Darwinian population will have these properties plus others. The behavior of a system is determined by the particular forms that variation, heredity, and fitness differences take, along with other features of the population. A particular parent (or pair of parents) cannot make it the case that their offspring's characteristics correlate or covary with theirs, for example. Covariance is a population-wide matter. All the parents can do is produce or influence their offspring, making them similar to the parents, in some ways at least. The evolutionary consequences of that relation depend on the population in which parent and offspring are embedded. Not all Darwinian populations need be changing, and a Darwinian population can change via non-Darwinian processes. But evolution by natural selection in a general sense is the large category of change *due to* variation, heredity, and reproductive differences, in some particular manifestation of those features and in conjunction with other factors. This approach combines the ingredients seen in the classical approach with a structural feature found in replicator views: a set-up or configuration is described first, and a range of models, which will often include idealizations, then describe how these systems behave.

The account above is permissive with respect to which collections of things can count as populations. The requirement expressed as "causal connection" is intended as shorthand for being located in a common network of causal interaction—with each other, with environmental conditions, or both. This is intended to be a weak requirement, allowing indirect connections. Some accounts of Darwinism apply a stringent standard here. Others impose no restrictions at all (though perhaps they are supposed to be present in the background). When criteria for membership in the same population are made explicit, they tend to require that the entities must be of the same species, and must occupy the same locality or not be too physically disconnected (eg., Millstein 2006, drawing on Futuyma 1986). Here I include a minimal requirement of the second kind, and none of the first kind. This will be discussed again in the next chapter.

The term "Darwinian individual" will be used for any member of a Darwinian population. Further, the idea of a Darwinian population is not set up in a way that is specific to some particular trait. Breaking organisms into traits is treated as something that comes later in an analysis. What is recognized first is the collection of individuals with all their features, both repeatable and unique.

I have followed the lead of both the approaches discussed in implicitly denying that evolution by natural selection can occur without reproduction. What is excluded is the idea that change due to mere differences in survival rate, in the absence of reproduction, count as evolution by natural selection. Traditional summaries have often handled this question awkwardly. There are two cases to consider: cases of differential survival without reproduction in things which *can* reproduce, and cases of sorting or culling in things which cannot reproduce at all. In the former case, the question is whether change due to differential survival is enough on its own, or whether it is just one stage in a larger process that amounts to Darwinian change. That seems to be primarily a verbal question, but I will follow the classical and replicator traditions in saying it is not enough on its own. There is a lot of attention to differential survival and "viability" in formal evolutionary models, but that is in a context where reproduction is assumed as part of the background. In the latter case, we are being asked to extend the Darwinian description to cases that lack a central feature of the phenomena that the theory deals with. It is *possible* to bend a partially Darwinian description around change in collections of things lacking reproduction, but this is a very artificial extension of the theory. Both those points seem to involve the management of words and little more; with ideas in a later chapter I will say something more substantial on the topic.

....

CHAPTER 3

VARIATION, SELECTION, AND ORIGINS

3.1. Beyond the Minimal Concept

[Paradigms and marginal cases; distribution and origin explanations; representing populations in an abstract space.]

The previous chapter developed a "minimal" concept of a Darwinian population and an associated category of change. These concepts are designed to be broad and permissive. Some "Darwinian" processes in the minimal sense are almost trivial. Others are able to build the most complex things we know of. That fact is one source of dissatisfaction with the standard three-part summaries (Sterelny and Griffiths 1999). A summary citing just variation, heredity, and fitness differences does not distinguish the events that gave us eyes and brains from a dull process of sorting fixed types within a population. That is true, but the minimal concept is supposed to be—as the name suggests—a starting point, the ground floor. So let us think next about a family of concepts describing evolutionary processes, some more demanding than others.

To mark out these relationships I will use a terminology that distinguishes *paradigm* Darwinian populations, Darwinian populations in the *minimal* sense, and *marginal* cases. The paradigms are the ones that have great scientific importance. These are the evolving populations in which significant novelty can emerge, the ones that give rise to complex and adapted structures. Paradigms need not produce things like us—the evolution of antibiotic resistance in bacteria is a paradigm case in this sense. The term "evolvable" is sometimes used for populations that have these features (Dawkins 1989, Kirschner and Gerhart 1998); they have the capacity to produce something new.

The *minimal* sense, again, picks out the broad category discussed in Chapter 2. The paradigm cases do count as minimal; they are a subset of those that meet the minimal criteria.

Lastly, *marginal* cases are those that do not clearly satisfy the minimal requirements, but only approximate them. So these are not the "dull" cases within the minimal category, but phenomena that have a *partially* Darwinian character. At the beginning of Chapter 2, I said that the aim of many summaries is to describe the "essential" features of natural selection. "Essential" here is understood in a low-key way, but the connotation was that of a definite category with fairly sharp boundaries. A Darwinian context, however, is one of transitions and intermediates. We should expect the process described in standard summaries to shade off into various kinds of marginal and partial cases.

In a vaguer way, I will sometimes also talk of "better" as opposed to "worse" Darwinian populations, and more and less "clear" cases. Here "better" (and so on) means *closer to the paradigm cases*; worse means closer to the marginals. This is the *only* meaning of "better" here. It is akin to the biologist's Greek prefix "eu" (good) in terms like "eubacteria" and "eukaryote," and it has no moral or normative loading. In this sense we can also talk of "better" cases of melanoma and genocide.

In getting a handle on the relations between the paradigm cases and the others, it will also be useful to make a distinction between two kinds of explanations in which Darwinian processes can figure, which I will call *distribution* explanations and *origin* explanations.

When we give a distribution explanation we assume the existence of a set of variants in a population, and explain why they have the distribution they do or why their distribution has changed. Some variants may be common, some rare. Some may have been lost from the population, having been present at an earlier time. A distribution explanation explains facts of that kind. An origin explanation, in contrast, is directed on the fact that a population has come to contain individuals of a particular kind *at all*. It does not matter how many there are, or which individuals are the ones bearing the characteristics in question. The point of the explanation is just to tell us how there came to be some rather than none. So now we are explaining the original appearance of the variants that are taken for granted when giving a distribution explanation.[1]

It is obvious that natural selection can be important in distribution explanations. It is less obvious that selection can have a crucial role in origin explanations as well. In a proximal or immediate sense, new variants appear in an evolutionary

---

[1] This terminology is modified from one due to Karen Neander (1995). Neander distinguished "creation" and "persistence" explanations in a debate with Elliott Sober (1984, 1995), over what natural selection can explain. Sober denied that natural selection can explain the characteristics of individuals. Neander argued that selection can figure in the explanation of the initial "creation" of an individual trait by reshaping the background against which mutations appear, and Sober seemed to be ruling this out.

I broaden Neander's two concepts, as well as rename them. Explaining distribution facts of all kinds is broader than explaining "persistence," and Neander tied her treatment of "creation" explanations to the idea that genes contain "programs" for traits, which is not needed. Aside from those details, I agree with the core of Neander's argument. (For refinement of the argument and a review of the debate see also Forber 2005.)

context via processes like mutation and recombination. And it can seem odd to say that selection, which has to do with sorting things that already exist, can somehow bring new things into existence. But natural selection can reshape a population in a way that makes a given variant *more likely to be produced* via the immediate sources of variation than it otherwise would be. Selection does this by making intermediate stages on the road to some new characteristic common rather than rare, thus increasing the number of ways in which a given mutational event (or similar) will suffice to produce the characteristic in question. Some kinds of novelty can be produced easily by an evolutionary process without this role for selection, but other kinds—complex and adapted structures—cannot.

The distinction between distribution and origin explanations will be used as one way to make precise the idea of a paradigm Darwinian process. When selection figures in an origin explanation, there is more going on than what is specified by the minimal concept. This is not to say that distribution explanations are shallow—many of them are not at all. And distribution explanations are *part* of any origin explanation in which selection plays a role. But the generation of complex novel structures is one mark of a significant Darwinian process.

This is also one of the most controversial parts of Darwinism. Creationists and other extreme anti-Darwinians will, in most cases, accept that selection has a role in distribution explanations of various kinds. But, they will insist, that is the end of the story; natural selection can do no more than sort pre-existing things. (See Pennock 2001.) The more striking Darwinian claim is that selection, in concert with random or undirected processes that generate variation, can produce something complex and new. As Stephen Jay Gould has put it (1976, 2002), the controversial claim is that natural selection *creates the fit*, as well as preserves them.

Up to this point I have handled the relations between paradigm cases and marginal ones (etc.) by treating these as a family of concepts. But a more informative thing to do is to *structure* that family, establishing relations between its elements. That will be done with the aid of a spatial form of representation. The idea is to pick a range of features that can be represented numerically. Each feature is associated with one dimension of a space. A population, in virtue of how it scores on each dimension at a time, occupies a point in the space. We can then ask whether, for example, the paradigm cases cluster in one part of the space and the marginal cases in another. The minimal criteria are supposed to pick out a large region of the space, covering the paradigm cases and shading into the marginal ones. Once we have this set of relations in place, however, the specific categories (paradigm, minimal, marginal) are to fade in significance. It is useful to have labels for important regions in the space, but the borders between them will be vague and partially arbitrary.

So through this chapter I will single out some features to use as dimensions, giving each a symbol (H, V, C, etc.) and later say something about how—squashing a lot of detail—these features might be represented with single numbers. This makes it possible to summarize some of the discussion in spatial terms.

3.2. Variation and Heredity

[*Reliable inheritance and the "error catastrophe"; particulate inheritance; the message of Fleeming Jenkin; replicators revisited.*]

First I will look at two factors whose importance is evident and much-discussed: the reliability of heredity and the supply of variation. This section will include some familiar messages rapidly covered, along with some rethinking and recasting.

Evolutionary processes with high-fidelity inheritance systems are different from those in which inheritance is very noisy and unreliable. Earlier my emphasis was on comparative measures of parent–offspring similarity. Regardless of how variable the population is, and how different parent and offspring tend to be, is the state of a parent predictive to some extent of the state of the offspring? That is what matters to whether fitness differences will produce any change at all. But now we are concerned with *absolute* reliability. When inheritance is very unreliable in absolute terms, the products of one round of evolution tend to be lost on the next, and also do not reappear later on. The result is little possibility for "cumulative" change, for evolutionary processes involving the successive addition of slight modifications to an existing structure.

This fact poses interesting problems for the explanation of the origin of life (Eigen and Schuster 1979, Ridley 2000). The "error catastrophe" encountered in models of early life arises from the fact that quite a lot of biological complexity is needed before an inheritance system can be reasonably reliable. But that complexity must itself evolve by natural selection. How is it possible to evolve the machinery needed for reliable inheritance itself? Coordinated and fine-grained chemical action is needed for high-fidelity heredity, but the enzymes needed for such action cannot evolve without high-fidelity heredity being present. A delicate boot-strapping process is required.

The story is also made complicated by the existence of sex. Talk of "reliability" of inheritance is simple when each individual has only one parent. But if there are two parents who differ with respect to a trait, an individual can only closely resemble one of them. The "input" to each reproductive event is a pair of individuals, though the output is a single individual. In such cases the offspring may resemble one parent only (as is seen in the case of an individual's own sex), may be a blend or average of the parental values, or may be something else altogether.

At this point some features of genetic mechanisms become important. As the story is often told, significant evolutionary processes rely on the inheritance of discrete elements—on "particulate" inheritance (Fisher 1930). Genes are passed down the generations with high reliability, but are constantly shuffled into new combinations. New genes are introduced through mutation, but this occurs against a background of mostly-faithful replication, retaining the finely-tuned products of earlier rounds of evolution. This fact is not immediately visible because, in a sexual population like ours, all the individuals contain unique combinations of genes. So populations like our own have low-fidelity inheritance at the level of organisms, due to sex, but high-fidelity inheritance at an underlying level of genes.

I assume that the heart of this account is right, but let us think further about some of the details. First, I resist the idea that we, as sexual organisms, are low-fidelity inheritors in the sense relevant here. It would be *possible* to have a situation where reliable genetic copying was accompanied by noise to the point of chaos in organism-level inheritance—producing an error catastrophe, in fact. But there is no problem in the human population of retaining the basic architecture that evolution has built. We, as organisms, have quite reliable inheritance, though less reliable and more complicated than it could be if we cloned ourselves or our population had less genetic variation. We are not clouds in the sky or dust storms in the desert, shambling transients knocked together by our genes, but a population in which characteristics show a mixture of reliable retention, loss and recovery through sex, and in *some* cases fortuitous and unrepeatable appearance. The mostly-faithful copying of stretches of genetic material is part of the mechanism by which these organism-level properties are achieved. It is by no means the whole mechanism; it would be possible, again, to have faithful low-level genetic copying and accurate synthesis of proteins without this giving rise to the whole-organism inheritance patterns recognizable in a population like ours.

Here is another way to make the point. It is common when emphasizing the role of "particulate" inheritance to note an objection that Fleeming Jenkin, a Scottish engineer, made against Darwin's theory in 1867. Jenkin noted that given Darwin's own assumptions about how traits are "blended" in sexual reproduction, evolution by natural selection would run out of steam. The point is usually made as one about the "loss of variation" that occurs with blending inheritance. Here is Jenkin's spectacularly racist illustration of his argument.

Suppose a white man to have been wrecked on an island inhabited by negroes, and to have established himself in friendly relations with a powerful tribe, whose customs he has learnt. Suppose him to possess the physical strength, energy, and ability of a dominant white race, and let the food and climate of the island suit his constitution; grant him every advantage which we can conceive a white to possess over the native; concede that in the struggle for existence his chance of a long life will be much superior to that of the native chiefs; yet from all these admissions, there does not follow the conclusion that, after a limited or unlimited number of generations, the inhabitants of the island will be white. Our shipwrecked hero would probably become king; he would kill a great many blacks in the struggle for existence; he would have a great many wives and children, while many of his subjects would live and die as bachelors; an insurance company would accept his life at perhaps one-tenth of the premium which they would exact from the most favored of the negroes. Our white's qualities would certainly tend very much to preserve him to good old age, and yet he would not suffice in any number of generations to turn his subjects' descendants white. It may be said that the white colour is not the cause of the superiority. True, but it may be used simply to bring before the senses the way in which qualities belonging to one individual in a large number must be gradually obliterated. In the first generation there will be some dozens of intelligent young mulattoes, much superior in average intelligence to the negroes. We might expect the throne for some generations to be occupied by a more or less yellow king; but can any one believe that the whole island will gradually acquire a white, or even a yellow population, or that the islanders would acquire the energy, courage, ingenuity, patience, self-control, endurance, in virtue of which qualities our hero killed so many of their ancestors, and begot so many children; those qualities, in fact, which the struggle for existence would select, if it could select anything? (1867: 155–6)

The focus of Jenkin's example here is not the loss of variation—he could have added that new variants arise constantly, without altering his main argument. What is bothering Jenkin is the loss of a particular favored phenotype, once it has been broken up by sex.[2] And then the primary reply by a modern Darwinian is not that there is some lower level at which something is inherited as a unit, and never lost despite all the sex. The primary reply is to note a surprising fact at the whole-organism level. This is the fact that just as it is possible for a "white" and "black" to produce a "mulatto" in a population with Mendelian inheritance, it is also possible for two "mulattoes" who mate together to produce a "white." That fact has a mechanistic *basis* in the features of genes and meiosis, but it is a fact about inheritance at the whole-organism level.

What I am doing is separating the patterns of inheritance in the Darwinian population we are considering—whatever that might be—from facts about the mechanistic basis for those patterns. This basis might, in some cases, involve the activities of *another* Darwinian population, at a lower level. But the patterns of evolution at level n are a consequence of the patterns of inheritance *at level n*, and the nature of the machinery at level $n - 1$ that gives rise to those patterns is a separate matter.

---

[2] See Boyd and Richerson (1985) for a formal model of blending inheritance with stable retention of variation. Jenkin also argues, before the quoted passage, that even a highly favored new trait is likely to be lost through bad luck. Here his example, in contrast to the quote, is the burrowing behavior of a hare.

In the light of this discussion, let us look again at replicators. I said in the previous chapter that replicators are not necessary for change by natural selection. I then noted that an advocate of the replicator view might say that though this is strictly true, replicators are needed for evolution to produce anything significant. Arguments of this kind are made by Maynard Smith and Szathmáry (1995). More specifically, they think that the crucial factor in producing significant evolution is the presence of *unlimited hereditary replicators*, ones that pass on minor variations and can also exist in an indefinitely large number of forms. They argue that few things on earth can play such a role. Molecules that can be replicated by template processes are one example, and certain kinds of culturally transmitted structures are another. Roughly speaking, the two kinds of unlimited hereditary replicators on earth are found in nucleic acids and human language.

For now I will focus on nucleic acids. There is no doubt that something like *replication* is a very important biological process. By replication I mean what Dawkins and others mean: the high-fidelity copying seen in molecules like DNA. If there is replica*tion*, then in at least a low-key sense, there must be replica*tors*, the things being replicated. But so far this just means that stretches of DNA are replicated. The stretches can be of various lengths, have no particular boundaries, and need not be discrete units in any evolutionarily important sense. To put it differently, what we have learned is that the replication of genetic *material* is an enormously important matter. Fragments of genetic material are reliably copied, and via sex they are mixed into new combinations. But someone who accepts the importance of these processes has not yet committed to replicators as evolutionary *units*—has not yet committed to there being any such things, let alone to their being important things. The nature of that commitment will be discussed in a later chapter.

So far in this section I have emphasized the need for reliability in inheritance. If heredity is *perfect*, however, there is no source of new variation. A population with no ongoing source of variation is not able to do much. In biological contexts this is not likely to be an issue, but it might be in certain kinds of cultural evolution, and the in-principle point is important. A paradigm Darwinian population has reliable inheritance mechanisms, but not too reliable.

I will finish this section by noting some further points about variation that have been discussed clearly by others (eg., Amundson 1989, Sterelny 2001, Gould 2002). Paradigm cases of evolution by natural selection require not just "variation," but variation of particular kinds. If all available variation involves huge jumps in a space of phenotypic possibilities, cumulative selection again is not possible. At least some of the variation must be slight in extent. It also needs to explore many different directions around the current state of the organisms; it cannot be too "biased". Recent discussions have also looked at another feature of the supply of variation. Lewontin (1985) suggested that significant adaptive evolution requires "quasi-independence" of traits; distinct features of an organism must not be so tightly coupled that a change to one implies changes to many. There should be various causal routes to variation in characteristic X, only some of which involve correlated changes to Y and Z. This is now often associated with a kind of "modularity" in the organization of an organism (Schlosser and Wagner 2004), and also with such features as the indirectness of gene action—the long and multiply modifiable causal chains that connect a gene to its phenotypic upshot (Kirschner and Gerhardt 1998).

There are lots of ways we might represent the "abundance" or evolutionary significance of variation. Here, though, I will only introduce a symbol for a very simple measure of variation in a population; I will use "V" for the amount of variation present at a time. And "H" will be used for a parameter marking the distinction between reliable and unreliable inheritance.

3.3. Origin Explanations and the Struggle for Life

[*The struggle for life; when collections form populations; sex and competition; origin explanations and absolute size.*]

A noticeable difference between Darwin's descriptions of natural selection and most modern summaries, including my "minimal concept," is that the recent ones do not refer to a "struggle for life." We have left behind talk of scarce resources and the production of more offspring than can possibly survive. Most presentations of the replicator framework also omit this idea. Why is that? Modern writers have generally thought that, once we think abstractly, a struggle for life is not essential to natural selection. For example, Lewontin, in the 1970 discussion in which his summary appears, claims that "competition between organisms for a resource in short supply is not integral to the argument. Natural selection occurs even when two bacterial strains are growing logarithmically in an excess of nutrient broth if they have different division times" (1970: 1; see also Lewens 2007: 60).

One can see the rationale for this element of abstraction, but it creates immediate puzzles. Continuing with Lewontin's example: if two strains of bacteria in one dish dividing at different rates make up a population undergoing change by natural selection, then what about two strains in different dishes? What if the second dish is across town?

So the bacteria example reveals a more general problem. We are used to thinking about natural selection in a context where the boundaries of the evolving population have already been marked out in some independent way.

Many discussions take this boundary-setting for granted.[3] In Chapter 2, when I introduced the minimal concept, I included a requirement that the entities in a Darwinian population be connected in a common causal network. Clearly this is saying very little. It may suffice to rule out some extreme cases, but it seems both vague and very inclusive. For example, by this standard, organisms of different species could be included in the same Darwinian population.

My approach is to continue with a permissive attitude to the minimal concept, but treat this problem as indicating another feature that distinguishes paradigm Darwinian populations from others. Paradigm Darwinian processes take place in populations that are "glued" in particular ways into natural units. This is compatible with the idea that the minimal Darwinian criteria can also be applied to collections that have little binding them together.

I will discuss two factors that have this binding role, without claiming that they are the whole story. The first is the one that would be immediately cited by many biologists: sex. One way in which individuals can be bound into a genuine population is by a pattern of sexual reproduction. The second factor is ecological. A number of ecological factors might be relevant here, but I will focus on one with special importance in the Darwinian context: reproductive competition. To motivate these criteria I will look more closely at origin explanations.

I start by revisiting an old problem about Darwinism. On one hand, it seems that natural selection is a wholly negative matter—a filter or a reaper. Selection can only favor some pre-existing entities over others, and cannot bring anything new into the world. On the other hand, natural selection is supposed to be essential to the way that Darwinism explains the origination of novel and complex biological structures without intelligent oversight. How can something with selection's "negative" character do that? One might reply that the answer is to look at selection over longer timescales. But how can putting together *lots* of subtractions or filterings give rise to this "positive" role?

To resolve the problem we need to think of selection as one element in a package which includes the immediate, proximal sources of new variation. Then we note what these sources of variation are able to bring about when selection is present, and when it is absent.

I will assume a framework in which only standard genetic mechanisms are operating. In such a context, the "immediate" sources of new variation are mutation, recombination, and the migration of bearers of novel traits from outside. Let us focus on mutation. Mutation produces new genetic variants; but it produces them *from* pre-existing genotypes, and introduces them *into* a context comprising other genetic and phenotypic features. Those two facts are the key to

---

[3] Exceptions include Sober and Lewontin (1982) and Darden and Cain (1989).

the "creative" role of selection; selection shapes populations in such a way that combinations of genes and traits that are otherwise very unlikely to arise via the immediate sources of variation, become much more likely to arise. It does this by changing the population-level background against which new mutations appear.[4] The literature contains many examples of this (see especially Dawkins 1986); here I will use a very simple one.

Building the genotype underlying the human eye involved bringing together many alleles, each of them originally the product of local mutational processes. (Here I ignore variation within humans with respect to eye genes.) Consider a collection of genetic material, Y, that has everything needed, as far as genes go, to make a human eye, except for one final mutation. So this background Y is such that *if* new mutation M arises against Y, it will finalize the evolution of the human eye. Initially, Y was rare in the population—it was the product of a single mutational event that produced Y from yet another precursor. Selection can make the appearance of the eye more *likely* by making background Y more *common*. This increases the number of independent "slots" in which a single key mutational event will give us the eye.

So selection affects the production of new traits by changing the background against which the "proximal" sources of variation operate. My example here concerned only a single mutational event, but this is a fragment of a case of cumulative selection, which usually extends over many such steps. This is how what looks like a "negative" process can be essential to the origination of eyes and brains. This is what Gould meant when he said, as quoted earlier in this chapter, that natural selection "creates the fit" as well as preserves them.

But let us look closer at this story. I said that the probability of the genetic basis for the eye arising is being raised by "selection." Selection was said to make a combination of traits more likely to appear by changing the array of backgrounds against which mutations arise. But strictly, what matters here is changing the *absolute* number of backgrounds that are of the right type for the next mutation to produce something important. The probability of the eye arising depends on the absolute number of appropriate "slots," not on how numerous these slots are in relation to inappropriate ones. Selection itself, however, is usually seen as something that changes *relative* numbers in a population; the absolute numbers drop out (Millstein 2006).

---

[4] Reisman (2005) shows that these arguments understate the case. The literature in this debate has assumed a very simple picture of mutation—including point mutations only, and with all transition probabilities treated as equal. But there are varieties of mutation in which the genetic background strongly affects the new configuations arising, such as inversions and gene duplication events. There are also various other fine-grained effects of an existing genetic sequence on mutation probabilities.

Given that what matters to these origin-events is absolute numbers, selection is only important in origin explanations when it affects absolute numbers. Usually it does. But the two can come apart. I said that if Y is one mutation away from the genetic requirements for the human eye, selection can make the eye more likely to arise by making Y more common. But in principle, we could make the eye more probable either by increasing the number of appropriate Y bearers relative to the inappropriate Y^*, or by increasing the numbers of *both* Y and Y^*. A global population increase is as good (a bit better, really) than selection in this context. A global population increase is usually a very temporary process, but that does not affect this initial point.

Similarly, there are ways in which selection for Y could *decrease* the chances of the eye arising. Suppose we have a population with cases of both Y and some Y^*, and selection is then introduced. Y does better than Y^*, but the form of selection is such that we have fewer cases of Y around than we had before. Y has become more common relative to Y^*, but less common absolutely. So introducing selection favoring Y can, in principle, reduce the probability of evolving the eye. Similarly, introducing a regime in which Y is selected *against* is consistent with *increasing* the number of instances of Y in absolute terms.

I now connect these points back to the "struggle for life." Selection involves competition. But competition can be understood in a weak or a strong sense. Suppose I have two offspring and you have one; I have more offspring than you. But it may or may not be the case that my having two rather than one prevented you from having two rather than one. There may or may not be a dependence between my absolute fitness and yours, so that a slot I fill in the next generation is a slot that you do not fill. When we have competition in the stronger sense, there is a causal dependence between how many offspring each individual has. Then if I am successful under selection, this implies not just higher numbers than you, but higher numbers than I would have had if selection had not been favoring me.

Competition between types is only important to origin explanations insofar as the "winner" is able to produce more absolute numbers than it would otherwise.[5] The key relationship here is not one that is usually the focus of evolutionary models, but the right concept is represented in models in ecology, often with the symbol a_{ij}. Suppose we have two populations undergoing growth. As each population increases, it may reduce the rate of its own further increase, due perhaps to crowding; this is "density-dependent" population growth. But each may also affect the growth rate of the other population. The symbol a_{ij} represents

---

[5] Nanay (2005) argues for a different connection between limited resources and the explanatory role of selection. He argues that when (and only when) there is limitation of resources, the success of type A in the generations before t creates more chances for a new individual of type A to be born at t. This argument does not involve the role of mutation, only persistence. See Stegmann (forthcoming) for criticism of the argument.

the effect on the growth of population i of adding one individual to population j. When a_{ij} and a_{ji} are both zero, there is no interaction between them. When both are one, adding one individual to j affects i to the same extent as adding one individual to i, and vice versa. Other values and asymmetries are also possible.

The a parameter usually represents the extent to which two populations are bound together by competition. I borrow this way of thinking, and the associated symbolism, for a within-population use here. I will treat a as an overall measure of the extent of reproductive competition between individuals in a population—the extent to which adding reproductive success to one individual reduces another's. Clearly not all comparisons within a population will be the same in this respect. (Mates are a special case, for example.) But I suggest that paradigm cases of evolution by natural selection occur in populations where a is in the vicinity of one.

Competition, in the strong sense, then has two roles in the discussion. First, it has a role in origin explanations. Suppose we pick two types of organism, in the same dish or in different parts of the world, that are not causally connected, and we track their relative reproductive rates. The relation between these rates does not matter to any origin explanation. When we pick two types competing for a single resource, the fact that one is doing well or badly relative to another does have consequences for how the first is doing in absolute terms, so their competition does matter to origin explanations.

Secondly, we can then see that competition is relevant to working out when a collection of reproducing things forms a definite Darwinian population. Sex gives us one answer to this question, but not all organisms are sexual. There are many puzzles about kinds and groupings in the case of asexual organisms. I don't think an appeal to competition resolves all of them, but it does help. Ecology in general is a "glueing" factor when we think about populations, but competition is an especially Darwinian glue (Ghiselin 1974, Templeton 1989).[6] There is also a further connection between the two glues discussed here: sex has an important role in origin explanations itself, as sex makes it possible for traits arising within two separate lineages to be brought together into a single organism (Muller 1932).

Vague talk of "causal connection" in a Darwinian population has now been made more precise, by spelling out the most relevant modes of interaction.

---

[6] The way in which ecological properties "bind" organisms into a population here is related to the concept of "demographic exchangability" used by Templeton (1989) as part of his "cohesion" account of species. A difference is that Templeton's concept, designed for the species problem, depends only on the "intrinsic ecological tolerances" of organisms and not on their location and actual pattern of interaction. The concept used here, in contrast, is supposed to have its application influenced by where organisms live, as well as what they are like (cf. the examples with dishes of bacteria).

Many of the paradigm cases will then be found within a species. But the basic Darwinian concepts can be applied more broadly. And they *must* be applicable more broadly, because much of life has a very uneasy relationship with the concept of a species (Franklin 2007, O'Malley and Dupré 2007). The criteria here show why some cases where species boundaries are unclear, or definitely straddled, can nonetheless behave as paradigms.

How attenuated must connectedness be before a collection of things cannot be regarded as forming a Darwinian population at all? Rather than a cut-off, there is a shading into marginal cases. Gradations will be found at the "better" end of the scale as well. It is sometimes unclear whether two partly isolated subpopulations should be regarded as a single evolving unit. A person might also wonder what happens to an analysis if they draw unusually broad or narrow population boundaries. When boundaries are drawn in a way that leaves out relevant members of a population, or includes organisms better seen as outsiders, the result is a collection that is not quite paradigmatic. This may show up, for example, when some crucial events seem to be occurring "off-stage." That is a sign that the boundaries have been drawn too narrowly.

3.4. Fitness and Intrinsic Character

[*Dependence of reproductive differences on intrinsic character; role in significant evolutionary processes; somatic cells.*]

The next factor introduced is more novel. This feature will be symbolized with S, and defined as the extent to which differences in reproductive output in a population depend on *intrinsic* features of the members of the population, as opposed to extrinsic ones.

The term "intrinsic" is controversial in philosophy, but the main idea is straightforward and I will assume it is useable. Intrinsic features of one object are those that do not depend on the existence and arrangement of other objects (Langton and Lewis 1998, Weatherson 2005). An example of an intrinsic property of an object is its chemical composition. *E*xtrinsic features do depend on the existence and layout of other objects; examples of extrinsic features are location, and being someone's cousin. So extrinsic properties are, roughly, relational ones, though some special cases make the term "extrinsic" better than "relational." Historically, intrinsic properties have often been considered more real or more natural than extrinsic ones. That idea I see as completely mistaken, but it can be rejected without rejecting the distinction itself.

S has several roles here. First, it contributes another piece to my account of paradigm Darwinian processes. Second, it will be part of the unorthodox treatment of the relation between selection and drift promised earlier. S will also eventually play a role in my discussion of levels of selection and evolutionary transitions. So this concept ties several issues together.

The first two of these roles can be introduced with a thought-experiment using a well-known example. Assume we have a population living in a deterministic world, so that all the births and deaths in the population have complete causes. Consequently, there will be a full (often convoluted) causal explanation for why any individual *a* had more offspring than *b*. Imagine collecting a total explanation for all the reproductive similarities and differences in a population, over some time interval. This explanation will include *some* role for the intrinsic features of the individuals, and some for the features of their environments. But all cases are not on a par, with respect to this breakdown of factors. The population might be one where the differences in reproduction are largely due to intrinsic features of the organisms. Alternatively, it might be one where those intrinsic features play little role, and extrinsic factors tend to determine the differences.

An example of a fact of the latter kind is as follows. Imagine two very similar individuals, one with a long and fecund reproductive career and the other struck by lightning before reproducing. Here a reproductive difference is due primarily to extrinsic factors—where two individuals were in relation to a lightning strike. And though I said the two individuals were "similar" just above, this meant they were *intrinsically* similar. They were different in extrinsic properties (which is why one died from the strike and one survived). This is the sort of thing represented by *S*; the difference we focus on is the difference between *a* out-reproducing *b* due to a lightning strike, and *a* out-reproducing *b* due to the fact that *b* had a genetic disease. In the language of fitness, *S* is the extent to which "realized" fitness differences in a population are tied to differences in intrinsic character. They will never be *totally* due to intrinsic character, but there is a difference in degree here. When differences in reproductive output depend mainly on things like location—on who is in the right place at the right time—and these extrinsic differences are not the result of other intrinsic features—we have a low *S*.

The idea is not that high-*S* cases take place in some sort of vacuum. The environment, together with the state of the population, determines which intrinsic features are worth having and which are not. The distinction, again, is between cases where *given* an environmental context, intrinsic features make the difference and those where they don't. Success due to camouflage or a good mating call count as high-*S*, for example, even though what counts as camouflage or a good mating call depends on context.

The symbol "*S*" is chosen for several reasons. It is a fitness-related symbol for biologists, but it is also reminiscent of the philosophical notion of *supervenience*. "A set of properties Y supervenes upon another set X just in case no two things can differ with respect to Y-properties without also differing with respect to their X-properties. In slogan form, 'there cannot be an Y-difference without a X-difference'" (McLaughlin and Bennett 2005, symbols changed). The distinction between intrinsic and extrinsic is often important in discussions of supervenience. Here, S measures the degree to which the difference-makers with respect to reproductive success *approximate* being supervenient on the intrinsic character of the members of the population. (If supervenience seems suspicious, just think of a dependence instead.)

S is another measure of the extent to which change in a population has a paradigmatically Darwinian character. This is seen in both extreme cases and more moderate ones. Taking the extreme cases first, the example involving lightning strikes is a standard illustration of change through "drift." But the role of S is also seen in non-extreme cases. Part of this has to do with interactions between S and the processes underlying variation and heredity. Location is again a good illustration. At first glance, it might seem that it is not possible to inherit one's location. If so, differences in location cannot constitute a source of variation in a Darwinian process. But it *is* possible to inherit your location, in the sense of inheritance that is relevant here (Odling-Smee et al. 2003, Mameli 2004). Parent and offspring often correlate with respect to their location. It is possible to inherit a high-fitness location; one tree can inherit the sunny side of the hill from another. But the significance of this inherited variation is limited. A population can near-literally "explore" a physical space, if location is heritable and is linked with fitness. It may move along gradients of environmental quality; it may climb hills, or settle around water. But to the extent that reproductive success is being determined by location *per se*, it is *not* being determined by the intrinsic features that individuals have. If extrinsic features are most of what matters to realized fitness—if intrinsic character is not very important—then other than this physical wandering, not much can happen.

Mutation and recombination enable a population to "search" (more metaphorically) a space of possible genetic and phenotypic properties. Mutation processes subtly change intrinsic character, and some sequences of mutational change give rise to whole new kinds of biological organization. But this only works when these heritable intrinsic differences do give rise to reproductive differences.

So the evolutionary role of intrinsic properties is different from that of extrinsic ones, except where the extrinsic ones are consequences of intrinsic properties (such as preferences). To say this, again, is not to deny that environmental properties are *important* in evolution, in both obvious and unobvious ways. The obvious ways are ecological. The unobvious ones include the role of environmental change in producing novel biological forms by "plastic" response on the part of organisms, and in revealing previously hidden genetic variation (Schlichting and Pigliucci 1998, West-Eberhardt 2003).

I will discuss one other illustration of the role of S here, which introduces themes to come. Consider cells within your body. Cells reproduce, by cell division. They also vary, and inherit many features when they divide. In addition, some cells reproduce at a greater rate than others, and some do not divide at all. We would expect, then, that when a cell acquires a feature that leads to a greater rate of division, that type of cell would become more common in your body. This would be the normal Darwinian expectation with asexual reproducing entities and reliable inheritance. But such a process can only happen if mutation or mutation-like processes can generate a variant with such capacities—some heritable property that makes it divide more rapidly. When they do arise, the result—in the extreme cases—is what we call cancer.

Once we think of our cells as a Darwinian population, it is surprising how rarely this happens. The rarity is partly because the cells in your body are derived from a uniform genotype (in the zygote), and have little time to explore the space of possibilities by mutation. It is partly because cells repair mutations when they can. Both these factors amount to a suppression of variation. But another factor is the suppression of S. Much of what determines whether a cell divides or not is how it is located in relation to other things, which inundate it with signals, control its nutrients, and interfere with it if it behaves abnormally. Cell fitness is not very closely tied to intrinsic character. And the most important manifestation of this is the enormous difference in fitness, over the longer term, between "germ-line" cells (the precursors to eggs and sperm) and the "somatic" cells everywhere else in your body. Cells outside the germ line can win a Darwinian contest for a few generations, but only for a few. Somatic cells, as is often said, are evolutionary dead-ends in principle. Particular germ-line cells may be dead-ends in fact, but all the humans on earth derive from germ-line cells that were not dead-ends. This huge reproductive difference between a successful germ-line cell and a somatic cell in the same person is not due to intrinsic character. It is primarily a matter of location. There is almost nothing that a cell in your liver can do, altering its intrinsic character, that can give it a longer-term evolutionary future.[7] If it mutates to a different gene sequence and multiplies, this might give it a few generations of descendants it would not otherwise have had. But that is as much as it can do.

Our bodies are made up of Darwinian populations, but these are populations whose evolutionary activities are not paradigmatically Darwinian. Part of the reason for that is the fact that the populations comprising these important parts of us have low S. This feature is an evolutionary product. In other organisms things are different.

---

[7] *Almost* nothing. See Burt and Trivers (2006), especially on canine venereal tumour. See also the new example of Tasmanian devil facial tumour (Pearse and Swift 2006). These are cancerous cell lines that have become contagious.

3.5. Continuity

[*C introduced; Wright, Gavrilets, and fitness landscapes; heat-shock proteins.*]

The final feature I will look at in this chapter has been widely discussed. I will symbolize it with *C*, which stands for "continuity," a term used by Lewontin (1985). An evolving population exhibits continuity when small changes to an organism's phenotype lead to small changes in its fitness.[8] Lewontin introduced this alongside a related condition, "quasi-independence," which was discussed in Section 3.2. We have quasi-independence when the features of an organism are able to vary fairly independently of each other. So quasi-independence is concerned not with fitness, but with which variations are possible. Continuity is concerned not with which variations are possible, but with the relations between changes to phenotype and changes to fitness. Lewontin suggested that powerful evolutionary processes require both continuity and quasi-independence.

Although this is not essential, I will discuss *C* in terms of the "fitness landscape." Sewall Wright (1932) introduced the idea of "landscapes" which represents the relations between organismic properties and fitness. We imagine properties of organisms (or sometimes, populations) represented in several dimensions, and fitness represented with another dimension, visualized as height. So mountains (if there are any) correspond to areas of high fitness, valleys to areas of low fitness. The landscape can be used to represent genetic properties and their fitnesses, or phenotypic properties. Either way, we have a "smooth" landscape when similar organismic properties are associated with similar fitness values. Variation in fitness appears then as smooth ascent to a mountain, or a number of rolling hills. We have a "rugged" landscape when similar organismic properties are associated with very different fitness values, yielding a landscape of jagged spikes and steep-sided pits.

This landscape metaphor has been both controversial and fruitful. (For a review, see Pigliucci and Kaplan 2007). In this book, inheritance is usually treated in a way that does not assume the presence of genes. So a "fitness landscape" describes a relationship between individual characteristics in general, not necessarily genetic characteristics, and fitness. The shape of a fitness landscape always depends on what is next to what—on what count as "nearby" possible states. For a landscape to *have* a definite shape, there have to be non-arbitrary facts about relative closeness and distance of this kind. This assumption is a strong one when all individual characteristics, not just genetic ones, are included.

---

[8] Here I simplify Lewontin's idea. He said we have continuity when small changes to an organism's phenotype lead to small changes in its functioning and relations with its environment, which should imply small changes to fitness.

But when this assumption does make sense, smoothness of fitness landscape corresponds to C.

There is a standard story about the importance of smoothness in fitness landscapes. An evolutionary process with good sources of variation and high-fidelity inheritance will tend to climb hills in the landscape. Populations will not tend to sit exactly on top of a peak, but form a cloud around and on it. The peak ascended will not necessarily be the highest one in the landscape, but a local one that first attracts the population. In a rugged landscape, the peak ascended will usually not be an especially high one, as most points in the landscape are close to low peaks, separated from higher ones by valleys. Without special mechanisms operating (a finely-tuned mix of selection, migration, and drift), a population cannot traverse a valley to reach other peaks.

This standard story is now controversial in several ways. Gavrilets (2004) has argued that the familiar image of an evolutionary "problem" of populations stuck forlornly on low peaks, requiring implausible and delicate machines to shift them, is an artifact of thinking about variation in a low-dimensional way. If we think realistically about the multitude of dimensions of variation in actual organisms, the alleged "peaks" in a fitness landscape will tend to be linked by bridging ridges. So Gavrilets is not disagreeing with the possibility in principle of an evolutionary obstacle posed by rugged landscapes; he is disagreeing with the claim that this problem will often be actualized. The use I make of C in this book does not require taking sides on debates of that kind. Gavrilets and others would agree that *if* a landscape has many truly isolated peaks, significant adaptive evolution is more difficult.

Talk of a "landscape" also suggests a kind of fixity in the situation faced by a population. That has been another source of controversy. If the idea is to be applied carefully, it has to include the idea that as organisms evolve, they can also change the landscape on which the population moves. For some (though not me), that makes talk of fitness landscapes permanently misleading. Certainly it means the idea should be used with caution.

C, as understood here, does not have to be understood via the metaphor of a landscape. I intend it as a rough measure of the overall extent to which similar organisms in a population have similar fitness. And the fact that evolution changes a population's fitness landscape is useful in introducing one of my main emphases. C is not a fixed feature of life for an organism of a certain kind in an environment of a certain kind, but changes as the population evolves. A good illustration of this is provided by recent work on "heat-shock" proteins in various organisms, especially a protein called Hsp90.

Heat-shock proteins assist a cell in the production of a normal protein from a gene despite adverse temperature conditions. So they are a buffer against environmental variation. It turns out, however, that they also act as a buffer against the effects of some kinds of genetic variation. When normal heat-shock proteins are absent, in model organisms such as fruit flies or mustard weed, a range of normally invisible genetic variation is unmasked in the form of deviant phenotypes (Rutherford 2000). There are mutations that do not have dramatic consequences in the presence of heat-shock proteins, that do have dramatic consequences without them. So if we take a total fly genotype that *can* produce a normal-looking fly in the absence of heat-shock proteins, this genotype is surrounded by many slight variants that cannot produce a normal-looking fly in those conditions. Using the landscape metaphor, when heat-shock proteins are absent the genotypes that produce a normal fly are surrounded by many deep holes. When heat-shock proteins are present, the holes are smoothed over.

So when cellular tools such as heat-shock proteins evolve, they raise the value of C. There are fewer low-fitness organisms that are small variants on high-fitness ones. When such devices are lost, C is reduced. C is not a fixed fact of life, but an evolutionarily tunable feature.

3.6. Selection and Drift Reconsidered

[*Drift as a force or a reflection of ignorance; drift as low S and low C; reconstruction of the role of population size.*]

In this section I will use the ideas outlined above to put a new slant on the relation between selection and "drift." This is the discussion that was promised in the previous chapter, when I sketched a rough distinction between the two but said that more was to come.

Drift, once again, is supposed to be evolutionary change that occurs randomly, by chance, or by accident. It is often treated in evolutionary theory as a factor distinct from natural selection. In selection, change occurs not by accident but as a consequence of advantages that some organisms have over others.

I will divide views about this distinction into two families, one much more heavily populated than the other. The first family of views, which I will treat as "standard," takes the situation above at least roughly at face value. Drift is treated as an evolutionary factor of its own, objectively distinct from selection. Some within this first family of views treat drift as a "force" (Sober 1984); others view it as a cause though not really a force (Stephens 2004). More recently, it has been argued that drift must be understood in statistical rather than causal terms. Drift is a kind of "statistical error" in evolutionary processes (Walsh et al. 2002).

It might be thought that these exhaust the options, but the views above all share the idea that selection and drift are two distinct features of evolutionary processes—distinct evolutionary "factors," in a broad sense of this term. The second family of views, which I will call "unorthodox," denies that the distinction marks anything real in nature. Instead, "drift" must be regarded as no more than an informal label used to describe cases where the causes of reproductive differences seem quirky and inscrutable given our imperfect knowledge and zoomed-out vantage point. We talk of "selection" when we think there is an identifiable reason why some individuals in a population are out-reproducing others. To someone who could see all the causal details, there would be no difference, certainly not one in the "causes" or "forces" operating. (See Beatty (1984) for discussion and Rosenberg (1994) for a defense.)

The strongest argument for the unorthodox view has always been the intrinsic oddity of the standard view. Someone who believes that drift is a distinct evolutionary factor is not usually denying that all the births and deaths in a population have causes. The question of the status of drift is not taken to depend on whether nature is deterministic at the physical level. We might find some of the causal processes underlying births and deaths quirkier than others, but is that a basis to treat drift as a factor objectively distinct from selection?

The strongest argument for the standard view is based on the fact that there is a *theory* of how drift works. Drift is very important in small populations, less important in large ones. The relation between selection and drift can be quantified: if s measures the size of the fitness differences between (homozygous) genotypes at a genetic locus and N_e measures the population's "effective size," then drift is more important than selection when $4N_e s$ is much less than 1. Drift is even something we can *manipulate* (Reisman and Forber 2005). By manipulating the population size in a controlled selection experiment, we can reliably change the outcome. So someone who says that talk of "drift" merely reflects our ignorance has to contend with the fact that drift is at least real enough to be a manipulable part of the evolutionary process. By any reasonable standard, isn't that enough for drift to be a causal factor of its own?

I will suggest a different way of looking at the issue. Some clues are already on the table. In my discussion of S, I used an example involving a lightning strike. When one of two similar individuals does not reproduce because he is struck by lightning, that is the kind of thing involved in a low S. Death by lightning strike is also a standard illustration of drift.[9] When the *reds* reproduce more than the *blues* because *reds* are better camouflaged, that is selection. When *reds* reproduce more because all the *blues* get struck by lightning, it is drift.

That suggests an association between S and drift. But the role of extrinsic properties is not all that is conspicuous about the lightning case. And some cases of "drift" do not seem to involve a special role for extrinsic properties at all. Organisms can have their reproductive output affected by internal accident, as well as through external, lightning-like events. So let us introduce a role for C

---

[9] Scriven (1959) is the usual cited origin of this tradition, though he used a bomb.

as well. C, I said, is a matter of the smoothness of a fitness landscape. Usually we think of this as a relation between fitness and familiar intrinsic biological properties. But extrinsic properties (like location) can matter to fitness, and more so when S is low. So let's think of C—at least for a moment—as describing the relation between (realized) fitness and *everything* about an organism—internal structure, place of birth, history of movement from one location to another, who it interacts with, and so on. Then when an individual dies by lightning strike, we have a low-C phenomenon as well as a low-S phenomenon. If the individual had done things a bit differently that day, it would have been elsewhere when the lightning came down. Lightning is both external and capricious.

When we think about change due to drift, the clearest cases are those where a population changes through events that are both capricious and not much dependent on intrinsic properties. That is, the population over that period has very low C and very low S. I realize that saying this involves treating the relationships described by C in an *extremely* fine-grained way, a way that is very different from familiar talk about fitness landscapes. But if we take the ordinary phenomenon of ruggedness of landscape and extend it towards a sort of limiting case, while also including extrinsic properties along with the intrinsic ones and assuming low S, the evolutionary change that will result in such a system will look like drift.

What looks *most* like drift is low C and low S. But cases where only C is low—cases where tiny internal accidents lead to reproductive consequences— might also look like drift. And there is a yet more attenuated sense in which low S without low C can look like drift. Suppose one type of organism happens to live in the wet part of an environment and another type in the dry part, and the former do better. If the internal differences between the types played no significant part in the reproductive differences, this *might* be called "drift" even though the role of the environment is more robust. If you insist that this last case is not drift, that is fine. The clearest cases have low S and low C. And to some extent, low S may lead to low C. Once intrinsic differences are assumed to have little importance to reproductive differences, a population will be hostage to various kinds of chaos and accident.

In these paragraphs I have made claims about how the term "drift" is used. This does not mean that my focus is the word itself. I am trying to describe the real features of situations that prompt and motivate talk about drift. I am offering a replacement for the usual distinction; the idea is not that drift "can be explained in terms of" S and C. Drift, as usually understood, is quite a problematic category. This is especially because the familiar ways of talking about drift involve a two-way distinction. Change, it is said, may be due to both selection and drift together, but these are treated as two distinct factors. I say there are not two "distinct factors" here, but distinctions along the gradients of S and C. Those parameters have motivations that are independent of this treatment of drift, and their role in this book usually does not involve the extreme values relevant here. The treatment of drift is a bonus. Once we are thinking in terms of S and C for other reasons, we can note that change involving *extremely* low values of these parameters has a non-Darwinian character that looks like drift.

Before moving on I will revisit the best argument for the "standard" view. This argument holds that there is too much well-confirmed theory about drift, and too much possibility for manipulating it, for drift to be regarded as anything but a real causal factor. A version of this challenge applies to me as well. If drift is really just change with low S and low C, why does it have special importance in small populations, and why can it be mathematically described in particular ways?

I do not have the expertise to tackle the mathematical description of drift in full generality. But what I can do is show why one family of cases (the ones used above) lend themselves to a description in terms of sampling and its consequences. When we have an extremely low value of C, we have a situation with an almost chaotic character. Each total set of properties of an individual gives rise to a particular reproductive output, but a very slight modification of those features will have a different upshot. Building on an idea due to Poincaré, Strevens has shown how this feature (along with one other) can make a system amenable to description using a probabilistic framework (Poincaré 1905/1952, Strevens 1998). Assume then that we can think of lightning strikes occurring during some time interval as removing a random sample from the population. Given the rarity of lightning strikes, this will be a small sample. Small samples tend to be unrepresentative of the populations from which they are drawn—at least, less representative than large samples. So lightning removes a small and potentially unrepresentative sample from the population before breeding. So far, that will be true of both large populations and small ones. But if the population itself is small, then removing a small sample will tend to make a difference to its properties. If the population is large, removing a small sample has little effect. The only way to have an (immediate) effect on a large population by removing a sample is to remove a large sample. But large random samples tend to be representative of the populations from which they are drawn, so again there is little effect on the frequencies of traits. This is one reason why causal processes that have a kinship to sampling are important in small populations and not in large ones. And many drift-like processes can be thought about in terms of taking a sample from the population and doing something to it (deleting, preserving, magnifying ... .).

In sum: drift is not a "force" in the biological world, but it is not a mere reflection of our ignorance either. If drift is anything, it is very low S and C—change via normal causes but with a particular role for distinctions between organisms that are extremely fine-grained, and for external difference-makers.

3.7. A Darwinian Space

[*Spatial representation of H, C, and S; neglected possibilities; movement on the space.*]

In this chapter I have discussed a range of features of populations that relate to the distinctions between paradigm, minimal-sense, and marginal cases. Five of them have been given symbols. These are:

H Fidelity of heredity
V Abundance of variation
a Competitive interaction with respect to reproduction
C Continuity, or smoothness of fitness landscape
S Dependence of reproductive differences on intrinsic character

The list is obviously incomplete. There has been no discussion of the strength of selection (size of fitness differences *per se*), and hardly any discussion of population size or the time available for a Darwinian process to run. Those are obviously important. A less obvious and more controversial factor is population *structure*; division of a population into subgroups, or other forms of spatial organization (Wright 1932). Another is "niche construction," causal feedback between organisms' activities and the environments they must contend with (Odling-Smee et al. 2003). The distinctive contributions of sex have also not been much discussed. To some extent this is because I have picked features about which I have something to say. In other cases (population structure, niche construction) the omission comes from the fact that I am focusing on more coarse-grained distinctions.

Even with the omissions, we have a long list of factors. Some order can be imposed—at cost of further simplification—with the aid of a spatial framework. The illustration I will use is a space that represents the roles of H, C, and S, as seen in Figure 3.1.

The figure can be read as representing a "projection" of a high-dimensional space into three dimensions. This operation is easy to think about for the relations between spaces of two and three dimensions. Imagine a transparent 3D cube, containing visible points scattered through it. If you then imagine looking at the cube directly from one side, you will see a flat 2D plane, and a number of the original points may be indistinguishable, as they differ only with respect to the dimension whose different values you cannot see. In the figures in this book, information about more dimensions is lost. In most cases, the way to think about Figure 3.1 is to assume that everything in the graph has high values on various unseen dimensions, so we can focus on the difference-making role of a few key factors. (The location marked "human cells" is an exception, as the cells have low V as well as low S.)

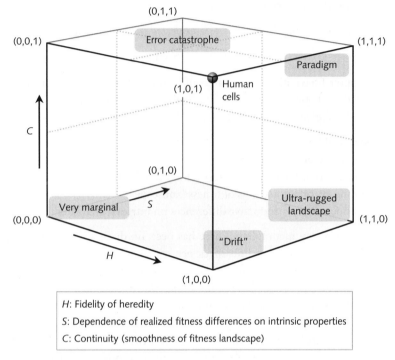

Figure 3.1: Spatial representation of cases in terms of (H, S, C).

I argued that high values of H, C, and S are associated with paradigm Darwinian processes. Starting from that $(1,1,1)$ corner, we see three ways in which a single feature can be changed to lead us away from the paradigms. If the fidelity of heredity degrades enough (low H), we reach the "error catastrophe." If, alternatively, the fitness landscape gets excessively rugged, so near-variants in phenotype have very different fitnesses, then adaptive evolution tends to stall (low C). Thirdly, a population can be in a situation where reproductive differences are uncoupled from differences in intrinsic character (low S). The label "human cells" is marked differently on the figure as it refers to the occupant of a region, not to the region itself.

We can also consider two moves away from the top corner. A combination of very low C and very low S—chaos plus the domination of extrinsic factors—yields the phenomena associated with drift. Two unoccupied corners correspond to high scores on other single dimensions; they do not seem especially significant, though I could be wrong. The $(0,0,0)$ corner is a region where all three features of the paradigms are lost—inheritance has collapsed, tiny variants have very different reproductive success, and those differences are not much dependent on intrinsic character.

How much sense does it make to score these features numerically? There are many ways it *might* be done, in each case; the problem is to do so informatively. *H* could be an average probability of the reappearance of parental traits in offspring. This assumes some way of counting traits, but it might be that similar numbers appear for many ways of counting. *V* might be measured with something like an entropy (which takes into account how many variants there are and how common they are) or in some cases a variance. *S* is less straightforward; here we might think in terms of an averaged measure of contingency that tracks how closely intrinsic variables are tied to realized fitness differences. *C* is naturally measured with something like the correlation between the fitness values of neighboring combinations. When pressure is put on these measures, my response is to suggest successive moves away from precision. Perhaps they can only be scored with respect to high, intermediate, and low values; perhaps only high and low. Some might admit of more gradations than others.

The space is intended as a heuristic device. Many important features of Darwinian populations cannot easily be measured with a single number; that does not make them less significant. But where the spatial approach is applicable, it has a number of benefits. First, it makes vivid the role of gradients and partial similarities between cases. Second, it gives us new ways of thinking about processes with a *non*-Darwinian character. To see this, imagine looking down various axes, away from the paradigms, into regions where parameters have values that are possible but rarely discussed. This is clearest with *H* and α (which is not pictured here). Parent and offspring may be well-correlated, may have no predictive relation between them, but also may be *anti*-correlated. Like can produce unlike (Haldane 1996). This possibility fits awkwardly into some discussions of natural selection, because it is a case where fitness differences do produce an immediate "response" in a population, but it is a response that takes the population away from such outcomes as an increase in the frequency or average value of a favored trait. Cultural transmission may sometimes have this character (as Haldane notes). Suppose that offspring tend to take on traits that are, within some domains, definite *functions of* their parental values, but not *similar* to the parental values. It was sometimes said that the offspring of 1960s hippies tended to be 1980s conservatives, and were so *because* of their parents' distinctive features.

Looking down the α axis, we see another neglected possibility. Competition is one way that reproductive rates can be tied together, but another is what Lewontin (1955) calls "facilitation," a *positive* impact of one type's reproduction on another's (negative α). Such positive relationships within a population will usually be type-specific and short-lived. They are common in symbiotic associations across different species, however. Again, the familiar Darwinian cases exist in a space also populated with converses and other possibilities.

So far I have made it appear that "more is better" with respect to each feature discussed; the paradigms are marked in the $(1,1,1)$ corner. This is only supposed to be very roughly true. Once we get beyond the coarse-grained contrasts made here, there is no reason to expect paradigm cases to be associated with higher values, as opposed to intermediate ones or complicated "balanced" combinations. This is conspicuous in the case of H. First, it was noted earlier that heredity can, in principle, be *too* high-fidelity, preventing the appearance of new variation. Second, in a sexual population like ours, the reliability of heredity is lower than it would be if we reproduced asexually via cloning. When there is sex, favored combinations of characteristics are constantly being broken up and rediscovered; individual quirks appear, are lost, and reappear in a new background. With asexual reproduction, a favored combination can continue indefinitely. This does not, however, affect the basic distinction between reliable inheritance and inheritance so noisy that cumulative evolution is impossible. So the paradigm region with respect to H should be pictured as spread over a larger region of that axis. Both sexual and asexual populations can be paradigms, of different kinds. They give rise to different kinds of evolutionary "search." Selection can figure in origin explanations in both cases.

The "paradigm" category, which I treat here in a simple and unitary way, could clearly be broken down further. The aim, again, is to mark an initial distinction between trivial Darwinian processes and important ones. Once we get past the coarse-grained distinctions made here, the question of where various kinds of significant cases are located in the space becomes an open question.

I will introduce one other heuristic role for the spatial treatment. As a population evolves, it not only changes the organisms that comprise it, but changes how it will evolve in the future. This can be visualized as movement in the space. Such movement is not always self-propelled; sometimes the evolution of one Darwinian population drives another. One example is the evolution of the vertebrate immune system. Here, whole-organism evolution has given rise to a suborganismal Darwinian process that is engineered to adapt effectively to the environment of viruses and bacteria that an organism confronts. Evolution at the whole-organism level has shaped parameters like S and H for the cellular components of the immune system, giving them the properties associated with a powerful Darwinian process. But it is also possible for one population to curtail or suppress another, with respect to its Darwinian properties, to move it away from paradigm status. Germ lines and other features of reproduction in complex organisms act to suppress or "de-Darwinize" the evolutionary activities of key parts of the system. The vertebrate immune system is engineered to perform a powerful evolutionary search, but somatic evolution is generally engineered to fail.

In the previous chapter I criticized the replicator view for its appeal to agential patterns of explanation, which interact with our psychology in particular ways. Some of the same message applies to the spatial tools used here. Once one invokes movement in a space, it can be tempting to think in terms of intrinsic directionality, or perhaps some sort of momentum created by earlier movement. There is no goal and no momentum.

CHAPTER 4

REPRODUCTION
AND INDIVIDUALITY

4.1. Reproduction as a Problem

[*The intuitive concept of reproduction; the link to individuality; problem cases.*]

Reproduction is at the center of Darwinism. This concept has been taken for granted so far. All talk about heritability and fitness requires that we know whether one organism is the parent of another. But the idea of reproduction is surrounded by uncertainties and puzzle cases. These problems, and their significance for Darwinism in general, are the topic of the next two chapters.

I will begin by sketching an informal or "intuitive" concept of reproduction, something close to common sense. That is not because we are bound to apply such a concept, or stay close to it. One focus of this chapter, in fact, will be tensions between our intuitive ways of thinking and what many of the biological phenomena are like. Our intuitive concept of reproduction has been shaped, understandably, by our experience with familiar cases. In some ways this concept guides us well, when thinking about evolution, and in other places it runs into trouble.

An initial analysis might be given by saying that reproduction involves the production of new individuals which are of the same kind as their parents. That is a shallow, but reasonable, beginning. More exactly, in a case of reproduction we have: (i) the production of a new individual, (ii) primarily via the causal role of some specific pre-existing individual(s), and where (iii) the "parent" individuals are of the same kind (broadly understood) as the new individual. Such a formulation leaves many questions unanswered, but does establish some key contrasts: reproduction can be contrasted with (i) growth of the same individual, (ii) appearance of a new individual without that individual having definite parents, and (iii) the production of waste and artifacts. These criteria are also abstract. In familiar cases the reproducing entities are organisms, but the criteria could also be applied to parts of organisms, collections of organisms, or things that are not biological at all.

All three of the criteria will turn out to cause trouble. A range of problem cases will be introduced in the next section, but we can start by recognizing two well-known families of difficulties:

(1) *Reproduction versus growth.* When is the production of new biological *material* the production of a new *individual*? The problem is most acute with plants and some "colonial" organisms, and in the absence of sex. Many plants produce new physiological units (things that *could* be seen as new organisms) from "runners" above ground or roots below. They are genetically identical (more or less) to the old plant. Botanists have debated for decades how to work out the fitness of plants that "reproduce" in this way.

(2) *Collective entities.* When do we have reproduction of a higher-level unit, as opposed to reproduction only of lower-level constituents that also come to have a particular organization? A herd of buffalo grows and then splits. Is that herd-level reproduction, or only buffalo-level reproduction? If we say it is only buffalo-level reproduction, then why isn't this reductionist attitude applied to ourselves, leading us to say that the production of a new human is merely a matter of cell-level reproduction along with a certain kind of organization of the cells? So here we encounter problems with the reduction of one set of entities to another, with how to think about levels of organization in nature, and again with individuality (Hull 1978).

After the survey of problem cases I will argue for a view of reproduction, as it figures in evolution, that is "permissive" in what it includes and has the gradient and multi-dimensional character seen in the analyses in Chapter 3. At different places in the tree of life, we find lots of different reproduction-*like* processes, lots of different ways in which new biological material is produced from old. This occurs for Darwinian reasons; the forms taken by the creation of new biological material are consequences of the contingencies of ecology and history on different parts of the tree. But the different ways in which new material is produced also have different consequences. The character of an evolutionary process is much affected by the kind of reproduction that the population exhibits.

4.2. A Reproductive Menagerie

[*Aspen, strawberry, oak; colonies and symbioses; chimeras and mosaics; the chimerically clonal grape; alternation of generations; formal reproduction.*]

This section will tour a number of hard cases, puzzles, and illuminating oddities. They are chosen because they put different kinds of pressure on our intuitive ways of thinking about reproduction. Some analysis will be given of each as we move along, but not much. The aim is to immerse our thinking in the diversity of cases first, and re-impose order later.

Aspen, strawberry, oak

Many organisms (various plants, animals, and fungi) create what look like new individuals by growing them directly from old ones. The new structure may then detach or stay attached. The new individual is genetically identical (roughly speaking—see below) to the old one. The organisms that do this also tend to reproduce sexually, though in some cases very rarely.

"Quaking aspen" trees (*Populus tremuloides*) are a famous example. What look like hundreds or thousands of distinct trees scattered across many acres will in fact be connected to each other by a common root system, from which they have all grown (Mitton and Grant 1996). In the terminology introduced by Harper (1977), in a case like this we have hundreds of distinct *ramets*, but a single *genet*, or genetic individual.

Similar phenomena are seen in violets and strawberries, which produce "runners" that give rise to new plants. In these cases the root systems are separate, produced afresh by the new ramet, and it is easy for the slender runner to be broken, resulting in complete physiological separateness. Separation can also be imposed; all the "Red Delicious" apple trees in the world are ramets derived from one apple tree that lived in Iowa (Pollan 2002). They are all parts of the same "clone."

Many marine invertebrates, such as corals, anemones, and ascidians, also do this sort of thing (Jackson and Coates 1986). In those cases we usually do not have "runners" that connect the new structures. The separateness of the new structure arises through simple fragmentation of the old. In some cases there are features suggesting that fragmentation is by evolutionary design, as opposed to mere happenstance.

A phenomenon that is sometimes put in the same category is *apomixis*, the asexual production of seeds by some plants, such as dandelions (Janzen 1977). Here the machinery of sexual reproduction is engaged, rather than mere growth, and the new physiological unit arises from a seed. But the new entity is, again, a "clone" of the parent. As in the aspens and violets, we can distinguish a large scattered genetic individual (Janzen used the term "evolutionary individual") from the physiologically independent units that comprise it. It will be useful to have a term for all entities that are prima facie new individual organisms, living things that are either physically separated from their "parents" or at least functioning largely independently, and regardless of their genetic properties and exact mode of production. I will use the term "physiological individual" (Cook 1980) for these entities. So the next question becomes whether the clonal production of a new physiological individual is ever a case of reproduction, or whether organism-level reproduction in the context of evolutionary theory is always the production of new genetic individuals.

A number of biologists, and also some philosophers who have looked at the problem, have claimed that, strictly speaking, only genets should be seen as individuals for evolutionary purposes: "genets are the fundamental units of populations upon which natural selection acts" (Jackson and Coates 1986: 8; see also Harper and Bell 1979: 30; Cook 1980).[1] The clonal production of new physiological individuals should be seen as growth. Such a view may seem strange, but there is a line of argument that can make it look quite natural.

Compare first the aspen grove, its ramets connected by a root system, with an oak tree. The oak has a trunk connecting its above-ground and below-ground branchings, while the aspen has a different shape and connections that are invisible. These differences in shape and visibility should not, it seems, affect how we count biological units. So the aspen and the oak seem to be in the same boat. Then turn to the violet. The runners are now narrower and flimsier, and may easily be broken, but the underlying processes by which ramets are produced are still the same. So Ariew and Lewontin, for example, conclude that: "If a tree is an individual then so is the collection of all the ramets of a violet" (2004: 360). This would seem to lead to the conclusion that ramet production is always growth. Surely that at least leaves the case of apomixis in a separate category, as here we have seeds initiating the life of the new physiological individuals? Janzen, in a memorable passage, argues for the opposite.

[T]he EI [evolutionary individual] dandelion is easily viewed as a very long-lived perennial organism. At any time, it is composed of parts that are moving around ("seeds" produced by apomixis), growing (juvenile plants), dividing into new parts (flowering plants), and dying (all ages and morphs). Natural selection could just as well have produced an organism with all these parts in physiological contact, but in view of the type of resource on which the EI dandelion specializes, this alternative arrangement of parts is clearly optimal. ...

In effect, the EI dandelion is a very large tree with no investment in trunk, major branches, or perennial roots. (1977: 586–7)

Janzen applies the same principle to aphids, which cycle between asexual and sexual reproduction. From this point of view, fragmentation and scattering is *one strategy* for a large clone. The other option is to remain intact. Which one is chosen will depend on the ecology of the organisms in question, along with their developmental resources (Oborny and Kun 2002).

---

[1] Cook: "From an evolutionary perspective, however, the entire clone is a single individual that, like you or me, had a unique time of conception and will have a final day of death when its last remaining stem succumbs to age or accident" (1980: 91). For the very interesting history of thought about these issues (Malphigi, Goethe, Erasmus Darwin …), see White (1979).

Colonies and symbioses

The other family of problems listed at the beginning of the chapter concerns collective entities. In one sense, of course, *everything* biological is a "collective" entity, as it is made up of parts. The important cases are those where at least some of the parts have the capacity to reproduce, and can reproduce largely via their own resources rather than via the coordinated activity of the whole. That is how "collective" will be understood here. The problems with collective entities again derive from the relation between reproduction and individuality. In the cases discussed just above, the question was when do we have production of a *new* individual as opposed to continuation of the old one. In the next set of cases, the problem is whether the entity produced is a genuine biological individual *at all*. I will illustrate the problem with two kinds of cases: colonial organisms and symbiotic associations.

"Colonial" forms of organization feature groups of entities of the same kind living in physically connected groups but without elaborate division of labor, and often with the retention of some capacity to live independently. Sometimes the "parts" are single cells; sometimes they are multicellular entities. The green algae of the group that includes *Volvox* are colonies of the former kind (Kirk 1998). A collection of algal cells is produced by asexual reproduction but they stay attached to each other. They form, according to the species, a clump or a sphere, and swim to different depths via the coordinated action of the cells' flagella. Examples of colonies of the second kind, where the parts are multicellular entities in their own right, are corals and some hydrozoans such as the "Portuguese Man O' War." Cells are tightly integrated into polyps and other "zooids," and zooids are more loosely integrated into visible colonies.

Colonies of this sort shade off into multicellular organisms in their own right, at one end, and temporary social aggregations, at the other. Sponges, for example, are usually seen as having shuffled far enough in the direction of integration to count as organisms. Some cases are fairly tightly integrated when they exist, but temporary. "Slime molds" are structures often formed during the life-cycle of the amoeba *Dictyostelium discoideum*. When food is abundant, these amoeba live in the soil as independent cells. When food is short, they join together and form a slug containing up to hundreds of thousands of cells, which first crawls to a suitable location and then forms an upright "fruiting body" that disperses some of the cells as long-lived spores (Bonner 1959, Buss 1987).

Symbioses are associations where the partners are very different kinds of organism, often from different kingdoms. A classic example here is lichen, which are associations between fungi and various kinds of green algae (and/or cyanobacteria). Lichen are perched so exactly between being organisms in their own right and being associations of different organisms that reading a textbook description of them can be a disorienting affair. Lichen have a distinctive ecological role. They can live in the most brutal environments, expanding at rates of millimeters per year, and often pave the way for other plants. Their photosynthetic parts can usually be found living alone as well. The fungi generally do not live alone, though they can be induced to if they are given enough nutritional help. Reproduction can be by simple fragmentation or by the formation of specialized propagules that contain samples of both partners. But the fungi may also form *ascomata*, the large reproductive structures of ascomycete fungi, which produce fungal spores alone. The spores form new lichens if they meet the appropriate algae.

Corralling photosynthetic algae or bacteria as a food supply is also seen in various animals, such as corals and clams. In some cases the bacteria are transmitted in the animals' eggs.

The most widespread symbiosis of all, however, is the eukaryotic cell itself.[2] After over a hundred years of speculation on the matter, it is now well established that both mitochondria (the sites of respiration) and chloroplasts (the sites of photosynthesis, in eukaryotes with that capacity) are derived from formerly free-living bacteria (Margulis 1970). There are other candidates, including the nucleus itself, which are much more controversial, but the significance of the two well-established cases is enormous. The introduction of the precursors of mitochondria is thought to have occurred between 2.2 and 1.5 billion years ago. Then between 1.5 and 1.2 billion years ago, one lineage of mitochondria-bearing cells also acquired a cyanobacterial partner, leading to the origin of chloroplasts in green algae and then plants. It is more accurate to call these cases *former* symbioses, as both mitochondria and chloroplasts have lost most of their genes to the nucleus of their containing cells. But they still have their own partially independent schedule of reproduction. Whereas the nuclear DNA in a diploid cell exists in two copies and is normally replicated once per cell cycle, the number of mitochondria in a cell is variable and they are constantly reproducing and dying. So the "reproduction of a cell" includes, as a component, the partially separate and ongoing reproduction of mitochondria, which are partitioned out to the daughter cells in a way that is thought to be more or less random (Burt and Trivers 2006).

As our knowledge grows, the importance of these "endosymbiotic" events looms ever larger, and the strangeness of the story grows as well. It now appears that some eukaryotes acquired their chloroplasts by engulfing another eukaryote. And in the case of some dinoflagellates, they came by *their* chloroplasts by engulfing some of those "secondary" engulfers. To finish this dizzying sequence,

---

[2] In this section I make extensive use of an excellent review of endosymbiont biology, and the history of this idea, in Kutschera and Niklas (2005). For the case of mitochondria I also draw on Lane (2005) and Burt and Trivers (2006).

some reef-forming corals contain dinoflagellates within their cells, alongside other passengers—some of the cyanobacteria whose photosynthetic abilities started the whole story off.

When we focus on these "tight" symbioses, from the lichen case to the eukaryotic cell, it can appear that the only natural attitude is a liberal one, in which collective entities can easily have their "own" reproductive capacities, over and above those of their parts. The fungi reproduce, the algae reproduce, and the lichen does as well. But symbioses come in all degrees of tightness. The corals have landed us back in the sea, so I will illustrate the other end of the spectrum with another marine case. At least ten species of gobies (small fish) live in symbiotic associations with individual shrimp, in small holes dug in the sand. In many cases the pair can be seen poking their heads out of the hole together. Here we have shrimp-level and fish-level reproduction, and it seems a stretch to talk of reproduction of the collective. But anyone who wonders whether their own domestic situation is a bit odd or implausible might take heart.

Chimeras and mosaics

We are used to thinking of individual organisms as both genetically *unique* and genetically *uniform* (Santelices 1999). Uniqueness raised problems above; now we look at uniformity. I will begin with a spectacular example. When marmosets give birth, it is usually to fraternal (dizygotic) twins. But these are not ordinary twins (Benirshke et al. 1962, Haig 1999, Ross et al. 2007). During pregnancy, links are usually established between the two placentas and, hence, the embryos. Cells are exchanged, and when each physiological individual is born its cells are a mixture of the genotypes produced by each fertilization event. So birth produces two genets and two physiological individuals, but the genets are spread across the two physiological individuals. There is no question that reproduction is occurring, and the question is how to think about the entities reproduced. If we follow the thinking of Janzen and others discussed above, with their emphasis on genetic identity, then when we think in evolutionary terms, the real individuals produced are the two spatially discontinuous genets. Haig (1999), who has given a detailed theoretical analysis of the case, is sympathetic to such a view.

Each physiological individual here—each marmoset-shaped object—is a *chimera,* a mixture of genetically different cells. The term "chimera" is sometimes used in a loose way for any organism that is a mix of genotypes, but I will follow the narrower usage, in which chimeras are distinguished from *mosaics.* Mosaics begin life with a uniform genotype, but become mixed as a consequence of mutations and other internal genetic changes, without (as in the marmosets) the bringing together of cells from different origins. (The exact relation between the phenomena will be revisited below.) The unwieldy term "intraorganismal genetic heterogeneity" (IGH) can be used to cover both (Pineda-Krch and Lehtilä 2004).

Chimerism is seen in spectacular form in the marmosets, but it is turning out to be much more common than had been thought. In humans, pregnancy induces a slight degree of chimerism in women that probably lasts for decades (Rinkevich 2004). Sometimes humans are massively chimeric because they are products of two fertilization events whose embryos merge and produce just a single baby. These cases are often discovered when the original embryos were of different sexes, so the result is an XX/XY chimera, which tends to be noticeable. There may be many cases like this where the merging embryos were the same sex, and the results are less conspicuous.[3]

Chimerism involves fusion; mosaicism involves internal change. And mosaicism, to various degrees, is routine. Any large and long-lived organism undergoes a constant turnover of cells. We might begin life as a genetically uniform zygote, but as cell lineages lengthen, they will genetically diverge. The replication of DNA is a highly reliable process, but mutation is not something that just characterizes the relations *between* individuals. In an organism like us, most genetic copying errors must occur within somatic cells, not the germ line, as a consequence of sheer numbers. Sometimes the results are markedly useful, as in the adaptation of the immune system; sometimes they are markedly bad, as in cancer. But divergence of genotypes across cell lineages is a fact of life, and one that becomes steadily greater to the extent that life is long.

At this point it is interesting to return to the discussion of ramets, runners, and trees from the first part of this section. Consider once again an old oak tree, with branches that diverged hundreds of years ago. Every branch on such a tree represents a separate evolutionary lineage. The tip of each branch extends by cell division in its "apical meristem," its growing point. Any mutation in a meristematic cell is passed on to its successor cells in the meristem as the branch lengthens. (The question of who exactly its "successor" cells are will be discussed below.) Further, because of the modular organization of trees, every branch on the oak is an independent site of sexual reproduction. The genetic material giving rise to pollen and ovules on one branch will be different from that on another; the most recent common ancestor of those cells may have lived hundreds of years ago. The same applies to distantly separated parts of an aspen "clone," except that the number of years separating two cells from their nearest common ancestor may now number many thousands. All through the discussion of ramets and genets above, as the reader may have noticed, I hedged or scare-quoted standard claims about the "genetic identity" of ramets. My reason was the inevitability of mosaicism.[4] Ramets may be very similar in their genotypes, but not (as it is often said) identical.

---

[3] Haig (personal communication) estimates that significant chimerism might have rates of the order of 1/1000 in humans.

[4] This claim for the *inevitability* of mosaicism is distinct from the more controversial claim that mosaicism in plants has an *adaptive* role. This is the "genetic mosaicism hypothesis" or

In organisms like us, whose sex cells come from a germ line that is "sequestered" early in life, mosaicism does not have the reproductive consequences that it has in plants. The divergence of branches on a tree—especially a tree with deep branchings, like an oak—is in a strong sense an evolutionary divergence (Whitham and Slobodchikoff 1981). Sex, if it happens, then takes the products of one divergent line and fuses them with the products of another.

The chimerically clonal grape

The problems posed by clonality, collectivity, and chimerism are brought together—with an air of trouble-making celebration—in the case of Pinot meunier, one of the three grapes traditionally used in the production of champagne.[5]

Grapes, like many commercially important plants, are often propagated by cuttings. The Pinots are an old group of varieties, and Pinot meunier has been handled this way for hundreds of years. So each Pinot meunier plant is a ramet, derived by growth (and removal) from earlier ramets. But now we look more closely at what ramets are like. In typical "dicot" plants like grapes, the "shoot apical meristems" (growing points at the tip of a shoot) have three cell layers, L1, L2, and L3. Each of these gives rise to different tissues in the plant, and gives rise to more meristematic cells of the same layer. Mutations can appear in any of the three layers. Each layer contains many cells, so a single mutation that arises may be lost, may take over the layer, or may persist with other cell types in its layer. Consequently, an individual branch may easily become a mosaic, by mutation in one cell layer that becomes established in that layer but not elsewhere. The material that will ultimately generate pollen and ovules comes from L2, so a mutation arising in L1 or L3 will not have consequences for sexual reproduction. But if the plant is propagated clonally, by runners or cuttings, then the mosaic state is preserved.

That is what happened with Pinot meunier. It is a close relative of Pinot noir, but has a mutation in the L1 layer that gives the plant somewhat different characteristics from the original—the plant is smaller, has different leaves, and the grapes ripen earlier. Each cutting brings with it all three layers and preserves the mix of genotypes. So if we trace back the lineages of cells, an L1 cell in a present-day Pinot meunier in France is more closely related to the L1 cells of a Pinot meunier in New Zealand than it is to the L2 cells next to it in the same plant.

GMH (Gill et al. 1995, Whitham and Slobodchikoff 1981), which claims that mosaicism makes trees phenotypically variable in a way that confers benefits when dealing with such threats as herbivores and pests. For discussion, see also Pineda-Krch and Lehtilä (2004) and the associated commentaries, especially Hutchings and Booth (2004).

[5] Here I draw on Boss and Thomas (2002), Franks et al. (2002) and Hocquigny et al. (2004). A few champagne houses use only Pinot noir and Chardonnay, while others see the addition of Pinot meunier as making a positive contribution.

And all of those cells have a most recent common ancestor cell in (most likely) medieval France.

Botanists who write about Pinot meunier call it a "chimera," like the marmosets. Unless this is the broad usage of the term, to say this requires that we treat each ramet as a new individual, starting life as a fusion, rather than as part of a big genet with mosaic structure. A number of other Pinot grapes have also turned out to be chimeras, along with one kind of Chardonnay. But this illustrates something that should—that *must*, to some extent—happen with many ramet-forming plants: the production of ongoing genetic mixtures, by initial mosaicism in a meristem that becomes chimerism once the branch acquires a life of its own.

Alternation of generations

The ramets, collectives, and chimeras discussed so far all have one reproductive feature that is straightforward and familiar: the new entity (the physiological individual) produced is clearly similar to the parent or parents. So we have no problem with the part of our intuitive conception of reproduction that says the new entity must be "of the same kind" as the parents. But in many organisms this is not so simple, owing to the *alternation of generations*. Here the intuitive notion of reproduction is disrupted in a new way. The parents of generation 1 produce entities that look very unlike them in generation 2, but when the members of generation 2 become parents, the results are similar to the organisms of generation 1.

This phenomenon is common—in some respects completely ubiquitous. Often it is not noticeable because the generation 2 organisms are not counted at all, or are treated as mere waystations. Dramatic cases are found in many ferns, including the "Filicale" ferns which include most of the familiar kinds. The fern-shaped plant or sporophyte is diploid (with two sets of chromosomes) and produces haploid spores (with one set) which disperse. When a spore germinates it grows into a new organism—a gametophyte—usually a flat heart-shaped structure which is often green and nutritionally independent. The gametophyte eventually produces gametes which, when they fuse, produce a diploid zygote. That zygote then grows up into the familiar fern-shaped sporophyte and the process continues.

This is a dramatic case because the two stages are separate, visible organisms, but it is not an especially strange one. Protists, primitive plants, fungi, and invertebrate organisms often go through elaborate sequences of states, increasing or decreasing their number of chromosomes, fusing or fragmenting, occupying very different environments. What we think of as the machinery of reproduction in large familiar organisms is often the evolutionarily-compressed remnant of a much stranger life-cycle; pollen grains are small and immature gametophytes. The cnidarians, which have furnished many examples for this section (corals, sea anemones, jellyfish) often go through two distinct life stages, the polyp and the medusa. In many cases the polyp is stationary and the medusa travels, but in the case of the "Portuguese Man O' War" the familiar floating and stinging entity is a combination of both, and—linking this section's problems once again—something that has long been used as an example of a colony which is not quite an organism in its own right (Huxley 1952, Gould 1985).

Back in Chapter 2 I introduced an imaginary scenario in which DNA was used to initiate an organism's life-cycle but was then dissolved, before being reconstituted in the making of a new generation of sex cells. This is a kind of "alternation of generations" scenario, though one at the molecular level. Such scenarios (not including DNA) have been used sometimes in discussions of the origin of life. The "hypercycle" model is one in which reliable reproduction of a kind happens, but in a cyclical structure: W produces X which produces Y ... which produces W. The hypercycle can be thought of as a reproducing entity with parts that are temporally rather than spatially organized (Eigen and Schuster 1979). The main message of alternation of generations does not require extreme cases, though. It is just the idea that reproduction-like phenomena seem not to require, in an evolutionary context, that parent and offspring be particularly "similar" things (Blute 2007). The road to the reliable re-creation of form can be more tortuous than that.

Formal reproduction

At the beginning of this section I said that the intuitive idea of reproduction includes a causal component; parents are causally responsible for the existence of the offspring. That is one feature that has not caused trouble so far. But we can separate out various aspects of the causal relations that were combined in the cases discussed above, and that are also combined in the most familiar kinds of reproduction. For example, we might distinguish the contribution of material to the offspring from the determination of structure or form. This sort of separation of reproductive roles has had a surprisingly important role in the history of biology, going back to Aristotle (who thought that fathers contribute no matter to their offspring, only a special kind of heat.)[6] The general idea of a distinct kind of causal role that involves contributing "form" has probably done more harm than good through this history (Oyama 1985), but there are some kinds of reproduction that can be usefully thought about in such terms. All these cases are found below the levels of organisms and cells—the significance of that fact

---

[6] James Lennox's summary from "Aristotle's Biology" in the *Stanford Encyclopedia of Philosophy* (2006): "The male contributes a source of movement or *dunamis* (power), which, as the argument unfolds, turns out to be a special sort of capacity to heat present in the semen's *pneuma* or air, which is part of its nature. ... The semen itself is merely a vehicle for delivering this warmth; the male makes no material contribution to the offspring."

will be discussed later. I will describe three examples: retroviruses, prions, and one kind of "jumping gene."

Retroviruses (including HIV) are viruses whose genetic material is RNA. On infecting a cell, they cause the copying of the viral genetic sequence into the cell's DNA. Later, the viral genes are transcribed back into RNA, and also induce the cell to generate proteins that will form the virus particle's coat. The "parent" virus particle is causally responsible for the production of a new virus particle very similar to it, but does not do this by contributing matter to the offspring.[7]

Prions have the same feature, in a very minimal form. A prion is a protein that is folded in a way different from the normal-functioning shape for that protein, and which is also able to induce other proteins of the same kind (same amino acid sequence or "primary structure") to lose their usual shape and take on the prion's strange folding (Prusiner 1998).[8] The results, which include "Mad Cow Disease," are medically disastrous. Once again, we can recognize a parent prion and an offspring prion. The parent is responsible for the offspring being the way it is, but that is only true with respect to one feature, the offspring's pattern of folding. The parent is not responsible for the material existence of the offspring, or for its amino acid sequence. There is one particular formal property that is "transmitted," and that is the extent of the causal relation between them.

Finally, a "LINE transposon" is one kind of genetic element which can multiply and move around within an organism's genome. The transposon codes for an mRNA molecule which is translated to produce a couple of proteins that immediately bind to the mRNA molecule itself. The proteins carry the mRNA back into the cell nucleus, cut the DNA on a chromosome somewhere (either randomly or somewhere specific), and reverse-transcribe the RNA back into the cell's genome. Thus a new copy of the genetic element is inserted, while retaining the old. Human genomes contain many copies of such elements.

Normally, the way DNA is replicated is for a double strand to split into two single strands, each of which becomes the template for a new double strand. So the "parent" molecule ends up with half its material in one offspring molecule and half in the other. The parent molecule contributes both part of the matter and also its organization to its offspring. A transposon engages in this ordinary kind of replication, but it can also cause new copies of itself to come into existence

---

[7] There is the possibility of some happenstance recycling of raw materials in the construction of a new virus particle, especially as retroviruses carry their own reverse transcriptase. But this is inessential to retroviral reproduction, and such material contributions could only be found in a small proportion of the progeny of a given particle. Here I use retroviruses as my primary example, but the claims made about formal parenting apply (more or less clearly) to various other kinds of virus as well.

[8] There is still some controversy around the views about prions sketched in this paragraph. I assume a "protein only" view of prions and prion diseases.

by the second, unusual route, in which it determines form (DNA sequence) but does not contribute matter.

In all these cases there is a chain of material *influence* linking parent and offspring, without the parent supplying a crucial piece of matter that initiates the new individual. This puts pressure on the causal element in the criteria for reproduction. What sort of causal influence, and how much of it, is required? Requirements of "faithful transmission of structure" were criticized earlier as too strong. Cases of formal reproduction reveal a different "gray area." In some cases it can seem that a "parent" is not really giving rise to a *new* entity, but reshaping or changing something that was already there. The prion example illustrates this—it is certainly very dubious as a case of reproduction. Other cases will be discussed later. Just as reproduction shades into growth, in the aspen, reproduction can shade into transformation as well.

4.3. Messages from the Menagerie

[*Genetic criteria and evolutionary individuals; significance of mosaicism; Griesemer and material overlap.*]

I will now start to draw conclusions from the cases discussed above. I begin with some critical points, in this section, and move towards a positive picture in the next.

I begin by looking at the role of genetic criteria for identity. It is common to think that these criteria have a deep theoretical role in this area; that in *some* sense, evolutionary theory tells us to count genetically identical things as parts of the same individual, no matter how odd that might initially look. That is the move often made by plant biologists when thinking about the vexed relation between reproduction and growth. One can see how it imposes order on the chaotic situation we encounter with runners and ramets. And even when we have what looks like the development of new plants from a seed, Janzen (1977) urged us to think about all the genetically identical dandelion plants deriving from a single fertilization event as parts of one big scattered object. The idea here is not that ramets and other asexually produced structures are additional *tokens* (instances) of a common genetic *type*. That is compatible with counting the entities as offspring. The idea is that asexually produced entities like ramets should be seen, for evolutionary purposes, as further parts of the same particular thing.

What is the underlying principle being applied here? Suppose it is something like this: reproduction requires the creation of a new biological individual, and a new biological individual must be genetically distinct from its parent(s). This principle could not be applied in a general way, however. Consider a dish of bacteria, dividing and competing. When a bacterium divides and does not mutate extensively in the process, this would not be counted as reproduction. *All* asexual "reproduction" would now count as growth, unless genetic change occurs in the process. Viruses would not be able to reproduce (except when they mutate or recombine along the way). Such a view also has strange intuitive consequences—monozygotic human twins would be regarded as two separately growing parts of one scattered individual—but what matters here is the evolutionary argument. Bacteria and viruses can clearly evolve. So the important sense of "reproduction" from an evolutionary point of view cannot require genetic novelty.

It is worth taking a moment to think further about what is going on here. The psychological pull of genetic criteria for identity is strong. And the most familiar cases of reproduction to us—human sexual reproduction—feature an obvious role for genetic novelty. Sex tends to makes reproduction clear, because the offspring cannot be a mere continuation of *both* parents. There is an intuitive sense in which a genetically novel individual is a fresh start, something new under the sun. We can also see how an appeal to genetic criteria becomes attractive in the case of plants and animals even though it is inapplicable to bacterial evolution. If we are looking at cell division, it is clear when something new has been produced even though sex is absent. But if we are dealing with many-celled entities, the relation between reproduction and growth becomes a problem. Then the idea that reproduction involves the creation of something genetically novel becomes attractive. Though this idea has an initial appeal when dealing with problem cases, it is not a good basis for a general account.

Further, all this talk of genetic "identity" and "novelty" involves an element of idealization, especially due to mosaicism. As in Chapter 2, I resist here the picture of genes and genotypes that emphasizes *identity* across cells or organisms, in favor of an emphasis on *similarity*. This might seem like a slight shift, but I think it generates a marked difference in how we think (Loxdale and Lushai 2003). Genetic similarity is one important kind of similarity. It comes in degrees, even within an organism. The divergence of branches in an old tree makes this phenomenon especially vivid, but the phenomenon of mosaicism applies to *some* extent to *all* multicellular organisms that have reasonably long lives, including you and me. We start our lives in a genetically uniform state, because we start from a single cell. But genetic change is ubiquitous in cell division, even though DNA repair mechanisms are arrayed against it. Our lineages of cells are slowly diverging; variation accumulates as the lineages lengthen. We are collections of cells that vary in their genotypes, very slightly or more substantially.[9] The oak's branchings are a spatial portrayal of something that is true of us all.

[9] Here are some calculations, using artificially sharpened-up numbers. Estimates of the overall rate of point mutations in eukaryotes, per mitotic division and per nucleotide, are around 10^{-9} or

The next set of ideas I will discuss is due to Jim Griesemer (2000, 2005). Griesemer has argued for some time that a concept of reproduction should be made central to foundational thinking about evolution, and has criticized the replicator approach. He has also started to develop a novel analysis of what reproduction is. His view is different from mine in several ways, and in this section I will say why I do not follow the same path.

For Griesemer, reproduction can be summarized as "multiplication with material overlap of propagules with developmental capacity" (Griesemer 2000: 74–5). The two distinctive features of this view are the requirement for *development* within the life of the individual and *material overlap*. I will focus first on the idea of material overlap. The idea is that offspring are "*made from parts of the parents, they are not merely similar objects made from wholly distinct materials*" (2000: 74, emphasis added). This, for Griesemer, is a key contrast with the replicator view. The concept of replication as discussed by Dawkins and Hull has a kind of formalist character, which Griesemer finds alien to the materialist emphasis of modern biology. Offspring are not merely similar things to their parents—in fact, need not be especially similar to them—but are materially derived from them. The production of organized and developing propagules by parents is how structure is made to reappear across generations.

I agree with Griesemer that material overlap is an important feature of many kinds of reproduction. But a concept of reproduction that *requires* material overlap is too narrow to work well in a foundational description of Darwinian processes. The reasons were introduced in the previous section, under the heading of *formal reproduction*. First, there are actual cases of entities that can undergo Darwinian evolution even though parents do not make a material contribution

10^{-10} (Drake et al. 1998, Ridley 2000). I will use a figure of 3×10^{-10} (Haag-Liautard et al. 2007, Otto, personal communication). Our diploid genome contains about 6×10^9 bases. Combining these numbers, we would then expect an average of about 1.8 point mutations per mitotic division (mother-cell to daughter-cell comparison). If two cells in your body are derived from a single cell 40 cell division events ago, for example, then assuming a simple "neutral model" with respect to cell fitness we would expect 144 point mutation differences between them (with the great majority of differences in non-coding regions).

The rate of mutation varies greatly across the genome. Microsatellite loci are especially mutable (with respect to insertions and deletions, not point mutations). Frumkin et al. (2005) construct a model using existing data that suggests a figure for microsatellite mutations in humans of the order of 50 new mutations per mitotic division, and a probability of a cell division event introducing no genetic differences that is extremely small. These highly mutable regions make it possible, in fact, to use phylogenetic methods to reconstruct the "cell tree" within an individual organism (see also Salipante and Horwitz 2006). Mitotic crossing-over is a further source of somatic change (Klekowski 1998, Otto and Hastings 1998).

So although it is common to say that almost all the cells in a human are "genetically identical," in fact none or almost none will be.

to their offspring. The most important case is provided by retroviruses, which clearly evolve (often to our detriment) and are among the paradigm cases. The parent–offspring relation found in the case of retroviruses is a clear one, with small variations reliably passed on, but is entirely "formal" in the sense of the previous section. Both the RNA and protein parts of the virus are manufactured by the infected cell. Except for possible accidents of recycling, no part of the offspring virus particle was once a material part of the parent.

Further, once we see the illustration provided by retroviruses, we see an in-principle point. There is nothing about the inner logic of Darwinism that requires material overlap as a feature of reproduction. Griesemer would be right to regard reproduction without material overlap as an oddity in the actual world—certainly I have produced only a limited list of cases. And one can see, also, that these special kinds of reproduction *depend* on the existence of entities that reproduce differently from them. But those observations do not affect the point that Darwinism itself does not require that parents make a material contribution in reproduction.

Griesemer's other requirement for reproduction is the capacity to "develop." Biological reproducers do not appear on the scene already bearing the capacity to reproduce again. They must achieve this capacity via change over their lifetime. My first point about this requirement is to say, again, that it understands the concept of reproduction too narrowly for the foundational role that is relevant here. Maybe viruses "develop," but it is pushing the concept pretty hard to say so. Yet the way one virus gives rise to another is sufficient for Darwinian evolution to occur. The idea of a requirement for individual development will, however, reappear in the next chapter.

4.4. Starting Afresh

[*Individuality relaxed; types and tokens; the concept of an organism.*]

I began this chapter by sketching an intuitive concept of reproduction, and then put pressure on it. Suppose someone was able to start afresh on the problem, aiming to construct a concept of reproduction specifically for evolutionary contexts. What would such a person come up with?

The resulting account would have a permissive or inclusive character, especially with respect to the link between reproduction and "individuality." There are lots of ways by which living things produce new *material*, and reshape old material into new things. Some of these look like clear cases of the creation of a new individual, and others look more dubious. But many are compatible with Darwinian change even when they are far from what we normally associate with reproduction.

The clonal production of ramets is, again, a good example. Once we see the connecting runner it may seem questionable to regard the ramet as a new individual. However, if a strawberry produces ramets that vary, that differ in their further ramet production, and that pass along their quirks to new ramets, then we do have the ingredients for Darwinian change. If ramet production is only a reproduction-*like* process, then reproduction-like processes are enough.

The ordinary term "reproduction" also has other connotations. The term goes most naturally with cases where a parent continues, as opposed (for example) to a case of fission in which there is no distinction between a continuing parent and a new arrival. Clearly fission is reproduction in the relevant sense, though. Both Maynard Smith (1988) and Gould (2002) use more specialized terms in this context—"multiplication" for Maynard Smith and "plurifaction" for Gould. I think the ordinary term "reproduction" is flexible enough, but those terms do both have one advantage: they imply a sense of reproduction that goes beyond mere replacement, or turnover. That feature will be discussed again later. One of the most obvious intuitive criteria for reproduction is physical separateness of the new entity. But the role of separation in the evolutionary context is not so clear, as ramets and other clonally produced entities can stay connected without much physiological exchange or dependence between them.

Other influences come from criteria for individuality and identity, criteria that bind objects into a unit and demarcate one unit from others. There is a long history of connecting the concept of an "individual" with that of an "organism" (Ghiselin 1974, Hull 1978, J. Wilson 1999, R. Wilson 2007). This connection can lead to trouble in the present context. First, Darwinian individuals in my sense need not even be close to being organisms. Genes, chromosomes, and other fragments of organisms can all form Darwinian populations. Even when thinking about organism-sized things with clear boundaries, rich thinking about "individuality" can intrude. Santelices, in an interesting review (1999), breaks what he regards as the standard concept of a "biological individual" into three criteria. In the most familiar cases, individuals are internally genetic homogeneous, genetic unique, and have "autonomy and physiological unity." These can be treated as three dimensions, each assessable at least for presence-versus-absence, yielding eight main categories. From my point of view, these criteria have very different roles. Part of this can be made clear by again making a "type/token" distinction. Reproduction is a matter of tokens, of instances or particular things. If one living thing produces another, it does not matter if they both fall into the same genetic *type*—if they are in some sense copies or duplicates. They are still distinct, countable things. "Autonomy and physiological unity" have to do with how a particular thing lives—how *that* thing lives, regardless of whether it is a duplicate of other things or not. One way to exist, to operate in the world, is as an organism, and physiological unity is relevant to whether an entity has that status. But not all Darwinian individuals have physiological unity—some do not have much in the way of physiology at all.

The link between "individuality" and reproduction is in some ways inevitable. Reproduction involves the creation of a new entity, and this will be a countable individual. But the right sense of "individual" to use here is a relaxed one. Two Darwinian individuals might be genetic duplicates (physical duplicates, in fact). One individual might be genetically heterogeneous. That is fine as long as we know who came from whom, and roughly where one begins and another ends. I said at the start of this chapter that our intuitive concept of reproduction has been shaped, naturally enough, by our experience with familiar cases. In some ways this concept guides us well, when thinking about Darwinian processes, and in other places it runs into trouble. The trouble has been especially conspicuous so far, but in the next chapter some of the concept's more useful contours will come to light.

CHAPTER 5

BOTTLENECKS, GERM LINES, AND QUEEN BEES

5.1. Three Categories

[Collective, simple, and scaffolded reproducers; compared to replicators.]

Reproduction is at the heart of Darwinian evolution, but modes of reproduction are diverse products of evolutionary history. The result, as we saw in the previous chapter, is a menagerie of reproductive processes found in different parts of the tree of life. There are lots of ways by which living things produce new *material*, and reshape old material into new things. These can be compatible with Darwinian change even when they are far from what we normally associate with reproduction.

The goal of this chapter is to impose some order. I will do that first by distinguishing three broad families, introducing terminology for each. The families are described by sketching partially idealized possibilities, which actual cases exemplify to various degrees. The families are not intended to cover all possible cases. The aim is to isolate three ways in which reproductive relationships can be part of a Darwinian process, each with different roles.

One of the three categories was introduced in the previous chapter. This is the category of *collective* entities, or rather, collective reproducers. These are reproducing entities with parts that themselves have the capacity to reproduce, where the parts do so largely through their own resources rather than through the coordinated activity of the whole. Not *all* the parts need to be able to do this for an entity to count as a collective, the requirement is that some can.

Examples in this category include the buffalo herd and other social groups, multicellular organisms like ourselves, symbiotic associations that are not too tight, and colonies. So collectives are cases in which a certain sort of reductionist description is at least *available*; a person might say: "this is not a case of reproduction by the collective; it is just lower-level reproduction plus a certain kind of organization of the results." This is not to say that such a reductive claim will be right, but it will be at least possible.

The second category is implicitly invoked by the first. If we have a collective whose parts can reproduce, those parts might themselves be collectives, or not.

But there cannot (except perhaps in strange imaginary cases) be collectives "all the way down." It will often be possible to isolate, within a biological system, the lowest-level entities that can reproduce largely "under their own steam"—or, more exactly, using their own machinery, in conjunction with external sources of energy and raw materials. I will call these *simple reproducers*. The paradigm here is a bacterial cell. Its reproduction is dependent on context—on the availability of nutrients, on appropriate temperature, and many other things. But (I hope this is not too metaphorical) it has the machinery of reproduction internal to it. And further, that machinery is not a collection of things that are able to reproduce under *their* own steam. Cell division is an activity of the whole cell.

Simple reproducers need not be the lowest-level reproducing entities in a hierarchy, however. A third category I will call *scaffolded reproducers*.[1] They might even be called *reproducees*, or at least many of them could. These are entities which get reproduced as part of the reproduction of some larger unit (a simple reproducer), or that are reproduced by some other entity. Their reproduction is dependent on an elaborate scaffolding of some kind that is external to them. However, these entities do have parent–offspring relationships, hence they form lineages or family trees.

Examples here include viruses and chromosomes. As part of cell division, a chromosome is copied; a new one is made from the old. The chromosome cannot do this with its own machinery, or even largely with its own machinery. It is more accurate to say that the chromosome is copied *by* the cell. Despite this, the new chromosome does have a particular parent chromosome. At least, a very newly formed chromosome has one parent chromosome; in organisms like us, there will then be *crossing-over*, which in effect gives a chromosome two parents. The examples of "formal" reproduction discussed in the previous chapter fall into the scaffolded category.

There is a sense in which hearts, lungs, and livers are also "reproduced" across generations, but that is a broader sense than the one I am using here. Each heart is not produced in a way that involves "parent hearts." Your parents did *have* hearts, of course, but their hearts were not causally involved in the appearance of your heart in the right kind of way to be parent-hearts. There is no way for newly arising quirks in their hearts to give rise to corresponding modifications in yours, for example. (This criterion is often used in discussions of replicators—they will be discussed again below.) Instead of reproduction linking the hearts themselves, your parents—as whole organisms—produced a zygote, which eventually grew a heart.

---

[1] My use of this term draws on Sterelny (2003), who uses it in the context of cognitive science to describe processes of learning that are scaffolded by instruction, artifacts, and the active shaping of the learning environment.

Returning to the case of chromosomes, some might say at this point that even though chromosomes need a cell's machinery to reproduce, the chromosomes—or rather, individual genes—are running the show, because they contain the "program" for the entire process. There is much to disagree with there (Godfrey-Smith 2007b), but my reply is that such claims are not relevant, even if true. In a material, mechanistic sense, the chromosome does not contain the machinery with which to reproduce, and that is the criterion used here.

So a simple reproducer can have reproducing parts, if those parts are scaffolded reproducers. If something has simple reproducers as parts, it is a collective. There can also be collectives of collectives.

Many actual cases fall outside and between these categories. We can distinguish two reasons for that. First, the categories are presented by sketching idealized possibilities. There are many real-world cases that do not exactly match any of them, but are much closer to one option than the others. Order is being imposed on an unmanageable menagerie, and this is being done in part via idealization. (The phrase "herding cats," used to describe tasks involving the management of wayward things, is especially appropriate here.) There are also cases that are "mixed" in a more important sense, because they are balanced between two categories, or are on a road from one to another. The eukaryotic cell is a former collective, and one that still has some features of a collective. Mitochondria, in different organisms, are at various locations on a road between simple and scaffolded.

Both collective and scaffolded reproducers may be sexual or asexual. The role of this distinction is less clear for simple reproducers, however. To say that something contains all or most of the machinery for its reproduction seems to imply that it needs no partner. The paradigm simple reproducers, cells, reproduce by dividing, though they also fuse. Could there be, in principle, a "simple" reproducer that needed a partner of the same kind? This is an awkward case for the taxonomy. One response would be to say that only the pair is a simple reproducer. Another would be to add a new category, though it seems odd for sex to generate a new category here when it subdivides the other categories. Yet another response is to see this as a further case that the taxonomy classifies as mixed or intermediate (between simple and scaffolded). This problem case is only a possible one, as actual cells divide under their own steam. That raises the question of whether it is an accident that life on earth features asexual simple reproducers in a central role, or whether there are deeper reasons why this is so.

The three roles described here are abstract, but it is clear which things they are abstracted from: organisms, cells, and genes. The replicator concept, discussed in earlier chapters, was originally an abstraction from the idea of a gene (specifically, the idea of an allele). It was later broadened by people like Maynard Smith and Szathmáry (1995), who were looking for a concept that would also apply to less gene-like things. They said that a replicator in a general sense is anything which "can arise only if there is a preexisting structure of the same kind in the vicinity" (1995: 41). This is a significant shift from what people like Dawkins and Hull had in mind. Understood literally, it includes multicellular organisms. So in a way, Maynard Smith and Szathmáry are expanding the replicator concept to cover the same sort of domain that I am trying to cover here. But if their goal was something like that, then the definition they gave was too broad. For example, all the enzymes that are used by a cell in the processes of gene expression are entities that can only arise if pre-existing structures of the same kind are in the vicinity. In my framework these do not count as reproducers, and the reason to exclude them is the fact that they do not form Darwinian populations. Each enzyme molecule does not have a "parent" enzyme molecule which gave rise to it. The enzyme has its amino acid sequence determined by some gene or genes, its raw materials contributed by food, and its construction achieved by many parts of the cell working together. Other enzymes are just parts of this machinery. Here the enzyme contrasts with a gene or virus particle. In these cases each individual is part of a network of parent–offspring relations. As a consequence, genes and viruses can exhibit fitness and heredity, whereas enzymes cannot.[2]

Scaffolded reproducers may also include things that only appear transiently, and do not persist to interact directly with their offspring, provided that the processes giving rise to them feature parent–offspring relations. Some viruses which lie dormant for long periods may be in this category. A more controversial case which could qualify in principle is a bird's nest (Bateson 1978, 2006, Sterelny et al. 1996). If each nest-building bird were to imprint on and faithfully copy the nest it grew up in, then nests could form a lineage of the right kind, even if the old nest disappears before its "offspring" nest is built. Here we might, in principle, find a feature that was lacking in the case of hearts: a tendency for new variations in a parental generation to reappear in the offspring, as a consequence of causal relations between the two. It is possible for there to be at least partial and attenuated parent–offspring relations in a case like this, with a fidelity of inheritance that is low.

One more thought before finishing this section: the category of simple reproducers is, of course, a pivotal one. On earth, cells are the distinctive occupants of this role, at least at the present time. Higher-level reproduction (reproduction of things like us, bee colonies, and buffalo herds) is elaborately organized cell division, combined with occasional cell fusion. Lower-level reproduction (reproduction of genes and chromosomes, especially) is organized, orchestrated, and made possible *by* cell division and cell fusion. If a Martian biologist came

---

[2] The same points apply to some treatments of replicator-like or reproducer-like units within the "developmental systems" movement (Oyama 1985, Griffiths and Gray 1994), and the "extended" concept of a replicator defended in Sterelny et al. (1996). These two options are discussed in Godfrey-Smith (2000).

down to earth, and started afresh on evolutionary theory using none of our usual concepts, I think that pair of facts would loom large. Cells occupy a special place, but they are often skipped over in foundational discussions of evolution, which tend to focus on the organism above and the gene below.

5.2. Bottlenecks, Germ Lines, and Integration

[*Parameters B, G, I; a spatial treatment; algae and bees*]

In this section I discuss one of the three categories—collective reproducers—in more detail. The approach taken follows the style seen in Chapter 3. I begin with a permissive attitude, welcoming all the unclear and peculiar cases under the heading of "reproduction." I then introduce features or parameters that distinguish the cases, and note their different roles.

The collective cases will be organized with three features. Two are specifically associated with reproduction, and the third has more general importance. The first will be symbolized with B, which stands for "bottleneck."[3] A bottleneck is a narrowing that marks the divide between generations. This narrowing is often extreme—to a single cell—but in principle is a matter of degree. So the degree of B is the "degree of bottleneckishness," the extent of the narrowing. This might be understood absolutely, or as a relation between adult and initial size. In the clearest cases we find both.

B matches, in an intuitive way, the idea of a "fresh start" at the beginning of life. It also has importance from the standpoint of evolutionary theory itself. One role will be introduced here, others later. The first is as follows. Because a bottleneck forces the process of growth and development to begin anew, an initially localized mutation can have a multitude of downstream effects. Part of this is a genetic matter; a single genetic change in the zygote ramifies into the genotype of every cell in the organism. But the underlying point is more general, and would apply even if genes did not exist. When a large organism starts life small and simple, it creates a window of opportunity for wholesale reorganization and change (Bonner 1974, Dawkins 1982a).

Thus the presence of a bottleneck has a link to the production of evolutionary novelty. In the language of Chapter 3, B has an important role in origin explanations. It affects the kind of variation that is available to an evolutionary process, making available variation that affects an organism's basic organization. This very fact makes individual development a precarious process, as most large-scale variations have bad effects. And in general, citing the effect of bottlenecks

---

[3] Here I re-use the letter "B," which is sometimes used to represent one of two competing types ("A versus **B**"), in other chapters. The bottleneck-related occurences of "B" are italicized, as seen with parameters G, I, H and so on, while the other uses are in boldface.

on variation is not to explain *why* bottlenecks are found. Their evolution will be discussed later. For now, the point simply concerns their effects.

The clearest cases where B is high are those where there are zygotes and other one-celled beginnings. These might be sexually or asexually produced. But B is intended here as gradient matter, not as a distinction which puts one-celled beginnings in one category and everything else in another. Considering again the cases with ramets and runners: the thinner the runner—especially in relation to what is to come—the less the new structure is a mere continuation of the old.

In many ferns, for example, a meristem at any time contains a single "initial" cell, which divides to produce a new initial plus a cell that contributes to the body of the plant. So when ferns produce ramets via underground rhizomes, as in bracken ferns, every cell in a ramet can have its ancestry traced back to a single lineage of cells in the rhizome. The bracken ramets themselves are not very large and elaborate, however. In plants such as the aspen, in contrast, a meristem has three layers (as discussed in Chapter 4) and contains, at any time, a dozen or more cells that play the role of "initials."[4] So the root projection that gives rise to a new aspen ramet is not so "narrow" in absolute terms. However, it is markedly smaller and less organized than the ramet it will produce. The resulting structure has a trunk, twigs, flowers, and leaves, with many sites of cell division. Aspen ramets have their own organization, regenerated via a cyclical process, and ramet production does feature a *partial* restart from something smaller and simpler. So the production of ramets in aspen can be seen as a case with an intermediate level of B.

The second parameter used in this section will be symbolized with G, which stands for *germ line*. G measures the degree of reproductive specialization of parts, in the sense involved in germ/soma distinctions and related phenomena. When G is high, many parts of a (mature) collective are *unable* to become the basis for a new collective of the same kind; they are, as far as reproduction of the collective goes, dead ends.

In mammals like us, for example, only a small proportion of cells can give rise to a new whole organism, those derived from cells "sequestered" early in development for the production of sex cells. The other "somatic" cells can reproduce at the cell level, but they cannot (at least naturally) give rise to a propagule that will become a new human. In contrast, any fragment of a sponge, if it breaks off, can grow up into a new sponge.

---

[4] Many details are presently unknown, and only some meristems have been studied. An entire meristem in a flowering plant contains perhaps a few hundred cells at a time, with a much smaller number (about a dozen, across the three layers) playing the role of "initials" which are likely to give rise to a long lineage of descendants. Cell numbers in root and shoot meristems are of the same order, and it is thought that there is little variation across flowering plant types. Initials can be supplanted and replaced, however, and there is a pool of neighboring cells that may be more or less likely to take on that role (Dumais, personal communication, Dumais and Kwiatkowska 2001).

The role of *G* is also illustrated by the celebrated case of eusocial insects, such as honey bees, in which the queen reproduces (along with the male "drones") and the female workers do not. This marks a distinction between cases where there is a group of insects who happen to live and interact together, and cases where the colony seems to count as a reproductive unit in its own right. This is not quite a "germ/soma" distinction in the normal sense. Worker bees, while a healthy queen is present, cannot usually reproduce at all.[5] They are dead-ends in a more immediate sense than somatic cell lines. I will sometimes use "germ" and "soma" in a broad way, though, to refer to parts of a collective that can, and cannot, give rise to a new collective through sexual or asexual reproduction.

There are various ways in which *G* might be measured. In some cases it might work to track the ratio of non-reproductive to reproductive units (Simpson, forthcoming). In other cases it might be best to use a set of rougher categories labeling germ/soma specialization as absent, partial, or present (Herron and Michod 2008). I will say more about this when I look at examples. The feature represented with *G* can also be contrasted with a different sense of "reproductive division of labor," seen in simple reproducers, like bacterial cells. Cell division is a whole-cell activity, in which various parts play distinct roles. There is no cell-level soma—dead-end material—but the division of labor within a cell certainly includes division of labor in the activities of reproduction.

Some animals have germ/soma distinctions and some do not, and there is variation also in how early in individual development the distinction is made (Buss 1987). Plants lack germ lines of the kind seen in us, but I will treat many plants, including the aspen, as intermediate with respect to *G*. This is because although a great many cells can give rise to a new ramet, *some* cells go down a developmental path that normally prevents their acting in the reproduction of whole new ramets. Leaves, for example, are dead-ends in many (not all) plants, as are various cells destined to make up internal vessels (Klekowski 1988: 165).

B and *G* will be discussed alongside another parameter, which is harder to define. That parameter is "integration" of the collective in an overall sense. This will be symbolized with "*I*," which I will take as a summary of such features as the extent of division of labor, the mutual dependence (loss of autonomy) of parts, and the maintenance of a boundary between a collective and what is outside it (Anderson and McShea 2001). These are notoriously elusive matters, but for much of my discussion only coarse-grained comparisons are needed.

G itself reflects one kind of division of labor; *I* reflects integration in ways other than a germ/soma distinction. My aim is to separate overall integration from the specific feature seen when a collective has soma-like or dead-end parts. A

---

[5] The assessment of *G* here (and below) is made complicated by asexual worker reproduction of males in some conditions in some eusocial bee species (Bourke 1988).

difference in I, holding G and B constant, is seen when we compare the splitting of the buffalo herd to fragmentation of a sponge. (Sponges can both have sex and fragment asexually.) The sponge is a more organized entity, with a division of labor beyond that found in the herd.

I see the overall role of B, G, and I as follows. In the case of collective entities, high values of all three parameters are associated with clearer or more definite cases of reproduction, as opposed to more marginal ones. When I say "clear," I mean that reproduction is less conflated with other things. Via B, reproduction is more clearly distinguished from growth. Via G and I, collective-level reproduction is more clearly distinguished from mere lower-level reproduction plus organization of the results. This is a place where some "intuitive" criteria for reproduction interact positively with what we learn from evolutionary theory itself; when B is high there is a "fresh start" in an intuitive sense, but also a sense that matters in evolutionary theory.

Sex is another feature that has the same role. Sexual reproduction establishes a new entity, as opposed to mere continuation of a parent. Sex does not handle all cases though. It helps with reproduction-versus-growth distinctions, but not with the problems involving low-I collectives like colonies and herds. There does not seem to be much difference, in this context, between a herd shedding a fragment which grows into a new herd, and shedding a fragment that unites with a fragment from another herd.

To make this handling of the contrasts clearer, I will again make use of a spatial framework. Figure 5.1 categorizes various phenomena discussed above and below. The aim of this first figure is to give a coarse-grained representation of many disparate cases. I suppose that when we put organisms as different as algae and ourselves on a single graph it is impossible to make fine distinctions with respect to B, G, and I, but possible and informative to make coarse ones. So here I make three-way distinctions on each dimension, between low, intermediate, and high values (corresponding to 0, $\frac{1}{2}$ and 1). What is represented in each case is a mode of reproduction. In some cases (like the sponge and aspen) an entity can engage in more than one mode.

With respect to B the distinction made is between the absence of any bottleneck (low), some significant narrowing (intermediate), and a minimally small (for example, one-celled) stage marking the start of the life-cycle (high). With respect to I, the distinction is between loose aggregations of entities capable of independent living (low), a level of integration seen in colonies and very simple organisms like sponges (intermediate), and the level seen in complex multicellular organisms (high). In the case of G, I distinguish cases where all lower-level units are capable (asexually or sexually) of giving rise to a new collective (low), cases of partial reproductive specialization (intermediate), and cases where there is a sharp distinction, established reasonably early in development, between germ

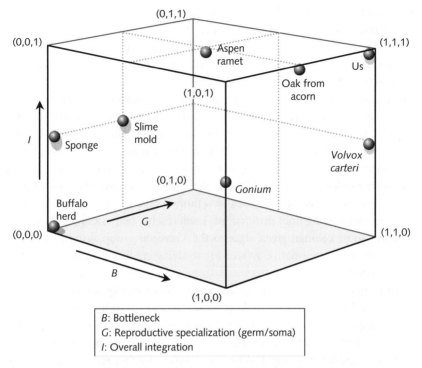

Figure 5.1: A space using three reproduction-related dimensions (B, G, I).

and soma (high). All the categorizations of cases in the figure are made in relation to a lower level in a biological hierarchy. In most cases the relevant lower level is the level of cells. The exception is the buffalo herd, in which the lower level is that of individual organisms.[6]

Working from the bottom left, the buffalo herd scores low on all three. A sponge reproducing by fragmentation (not through sex) differs from the herd only in I. A slime mold—here the reproducer is taken to be the fruiting body, making more fruiting bodies—has an intermediate level of integration, some reproductive specialization, and no bottleneck. A new fruiting body is formed by aggregation of many single-celled organisms; it does not grow by division from a small propagule. I treat seed plants like the aspen and oak as having a high degree of integration. The aspen is intermediate with respect to both B and G, as discussed above.

Gonium and *Volvox carteri* are colonial green algae—they will be discussed again in a moment, and distinguished with respect to I. In Figure 5.1 they

---

[6] In the case of plants, many botanists also recognize the level of the "module" between cells and ramets (eg., Vuorisalo and Tuomi 1989a). This level will be discussed a little in the next chapter, but I do not include it here.

both count as intermediate in I, reproducing through a bottleneck, and are distinguished by the fact that the former has no reproductive specialization whereas the latter has a strict germ/soma distinction. That leaves the oak growing from an acorn, and ourselves, both multicellular organisms reproducing through bottlenecks, differing only in G.

As the figure illustrates, we might expect the three features to be correlated to various degrees. They are intended to be *logically* independent, however; any combination of low and high values is possible in principle. The area corresponding to a combination of high G and low B is unoccupied in the chart, for example, but it would be possible to have a collective with a sharp germ/soma distinction that did not reproduce through a bottleneck. Perhaps there are real cases of this kind.

This framework can also be used in a more focused way—choosing only a few cases and making finer distinctions. I will discuss two examples. Figure 5.2 compares some colonial green algae in the *Volvocine* group, a group which is often seen as an informative system for studying multicellularity (Kirk 1998, 2005, Michod et al. 2003).

These organisms grow in aquatic environments, especially ponds. A single cell divides repeatedly to produce a colony, which may be of various sizes and degrees of organization. The colonies swim using their members' flagella, migrating to shallow water to take advantage of sunlight during the day and collecting nutrients in the depths at night. When food is plentiful they reproduce asexually, forming new colonies inside the old from single initiating cells. The new colonies bud off or are released from inside the old colony. When food becomes scarce they enter a sexual cycle, producing "zygospores" which lie dormant until times are better. Here I consider only their asexual mode of reproduction.

Figure 5.2 zooms in on a slice of Figure 5.1 on the far right-hand side, assuming a high value of B. With respect to G, I make only the three-way distinction described above. Further distinctions are made with respect to I, though in an admittedly impressionistic way. These are distinctions within what Figure 5.1 treated as an intermediate value of I. The sketches of the colonies are based on Kirk (2005) and are not to scale.

Starting on the left, *Gonium* colonies consists of loosely organized, flat clumps of 8 to 16 cells. All cells function in swimming, and all can reproduce. *Eudorina* forms organized spheres of 32 cells, with a distinction between inside and outside, but with no reproductive specialization. In *Pleodorina*, in contrast, there is a partial reproductive division of labor. These colonies contain 64 or 128 cells. All cells start out with flagella, swimming, but some give up this somatic function and become reproductive. The other cells do not reproduce. So there is now intermediate G, but roughly the same level of overall integration of the colony as in *Eudorina*. Lastly there is *Volvox carteri*. The colonies now number 2^{12} cells, in a highly organized sphere, and have a sharp distinction between germ and soma.

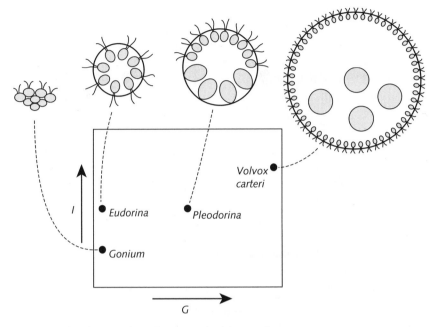

Figure 5.2: (G, I) comparisons for some colonial green algae.

Only a tiny percentage of cells are reproductive, and this role is allocated early. The vast majority of cells are dead-ends.

Figure 5.3 compares various kinds of bee colonies. The most famous bee colonies are the *eusocial* ones, with distinct reproductive and worker castes, and an elaborate division of labor. But bees also have several other kinds of social structure (and many species are entirely solitary).

The same three-way distinction is used with respect to G. The I axis now spans something like the entire range used on Figure 5.1, from loose aggregations to highly integrated and well-bounded collectives. The exact locations of cases with respect to I are impressionistic, however. It would also be difficult to align these cases with respect to the ones on Figure 5.1, especially for cases on the upper right, and I've not tried to do that. It is clearer to treat the bee comparisons only in relation to each other.

Working from left to right, the simplest bee social structure is usually referred to as *communal* (Michener 1974). Here a number of females make use of a common nest. Each reproduces, and can hence give rise not just to new bees but also (in part) to a new colony. Each female provisions its eggs independently. There may be some cooperation in defense, but there is no sharing of parental care or other division of labor. The size of colonies ranges from a pair to over a thousand. Cases of communal organization can be found in the small "sweat bees," so named

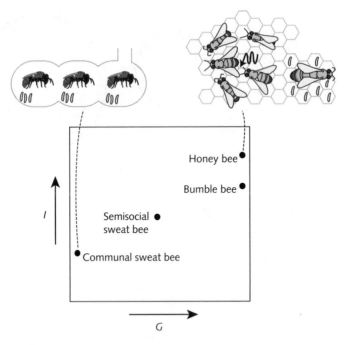

Figure 5.3: Bee colonies compared.

because of their liking for perspiration, and who are often a beautiful metallic green. (One example is *Agapostemon virescens*: Abrams and Eickwort 1981).

Communal organization shades at one end into local aggregations of "solitary" bees—cases where nests are close but not shared (E. Wilson 1971: 99). Here there is no collective at all. At the other end, communal organization shades into *quasisocial* organization. In these cases all the females are potentially reproductive, but there is some cooperation in caring for the brood. So quasisocial bees, if they were marked on the figure, would be higher in *I* but the same with respect to *G*. It seems to be controversial whether this category presently exists in nature (Wilson 1971, Crespi and Yanega 1995).

The intermediate case that I have marked instead is the *semisocial* level of organization (still Michener's terminology). Examples (with some controversy) can again be found within the sweat bees, notably *Augochloropsis sparsilis* (Michener and Lange 1958). These colonies contain a number of females of the same generation, who differentiate into a majority who both forage and lay eggs, and a smaller group who do not lay eggs at all. So there is some degree of reproductive specialization, but no differentiation into castes differing in size and shape. They cooperate in the provisioning of offspring (hence the higher *I*).

Then, on the right of the figure, there are colonies in which a single queen lays the vast majority of eggs, most females are non-reproductive workers, and these "castes" are determined early in each bee's life. Two examples are represented on the chart, honey bees (*Apis*) and bumblebees (*Bombus*). Honey bee colonies contain tens of thousands of individuals, with sharp divisions into castes and elaborate division of labor. Colony members engage in intricate communication, using the "waggle dance," to organize foraging, and chemical alarm signals. Bumblebee colonies, in contrast, have populations numbering in the hundreds, less differentiation between castes, less elaborate means of feeding the young, no dances or chemical alarms, and some internal aggression between individuals (Wilson 1971: 88). In some, though not all, bumblebee species the workers lay unfertilized eggs which grow up into males. This can happen in honey bee colonies too, but it not as normal (Bourke 1988). When bumblebee workers lay eggs we have a reduction in G. As part of my aim here is to show how I and G are biologically correlated but in principle distinct, the bumblebee species in the figure should be assumed to be ones in which only the queen lays eggs. A eusocial species in which workers lay a lot of eggs would be shifted to the left with respect to G.

Here I have only represented G and I for the bees, but there are interesting issues with B, the bottleneck feature, as well. In the communal case, collectives are formed by aggregating females (who need not be closely related: Kukuk and Sage 1994). So the communal bees are close to the (0, 0, 0) corner of Figure 5.1. In bumblebees, each colony is started by a single female (high B). Honey bees, in contrast, initiate colonies in swarms containing a queen and many accompanying workers. So the "propagule" is large, in both relative and absolute terms, but the workers in a new queen's swarm (an "after-swarm") are in some ways akin to the extra material in a large egg, produced by the mother colony. In time they are replaced by offspring of the queen.

Figure 5.3 represents present-day bees, but the distinctions made there are thought to correspond to one of two main evolutionary roads to eusociality. This is the *parasocial route*, in which nest-sharing and cooperation between females of a single generation leads, in time, to reproductive division of labor and elaborate social organization. The other path, the *subsocial route*, begins instead with mothers who remain in contact with their daughters. The parasocial route is thought to be specific to bees (though not to all of them), while other social insects like ants and termites are thought to have taken the subsocial route (Wilson 1971: 99). That is one reason why the semisocial category is of such interest—as a waystation on a road to full sociality. The path from lower left to top right in the *Volvox* figure (5.2) is also hypothesized to correspond, at least roughly, to an evolutionary road that was actually taken (Kirk 2005). So while the algae and the bees in these two figures are all extant organisms, the points they mark illustrate the dynamic interpretations of the spaces used in this book.

Here I used B, G, and I to handle collective entities. It would be good to apply a similar style of analysis to the other two kinds of reproducers, but these particular parameters do not seem so helpful in the other cases. A high value of I is almost inevitable in a simple reproducer, like a cell, and not needed in a scaffolded reproducer. Many scaffolded reproducers do not contain much "machinery" of their own at all. They are special parts of the machinery of a simple reproducer (chromosomes), or enter into Darwinian processes via the machinery contained in other things (viruses). B and G do not play much of a role here either. Thinking in a literal-minded way, it does not make much sense to even ask about B and G in the case of things like cells. But understood abstractly, we can ask whether cell division involves a narrowing and reduction followed by a rebuilding, and whether some parts of a cell are soma-like in their role, and the answer in each case is no. So high values of B and G are not necessary for definite reproduction in all cases. The account of collective reproduction here is intended to deal with a particular kind of "pressure" on the concept, the pressure arising from the existence of reproduction at a lower level. A different kind of pressure arises from questions about the boundaries of reproducing entities. That pressure, which arises for all three of my categories, has not been addressed in any detail here (see Griffiths and Gray 1994, Turner 2000). I have also left unresolved many problems falling under the "alternation of generations" heading in Chapter 4.

Earlier I discussed Griesemer's account of reproduction, which includes requirements of "material overlap" between generations, and development (Section 4.3). We can revisit that discussion now the simple/collective/scaffolded distinction is on the table. Neither material overlap nor development are needed in the scaffolded cases. Simple reproducers will generally reproduce with material overlap and development, it would seem, though perhaps this is not absolutely necessary. In the collective cases, the presence of a bottleneck does imply something like development. One way to look at the situation is like this: material overlap and development are characteristic of many reproducers. But once we have individuals of those kinds, the possibility arises for other entities to reproduce—and evolve—differently.

5.3. De-Darwinization

[*Subversion of higher level; suppression of lower level; B and G as de-Darwinizers of lower-level entities.*]

Chapter 3 used five parameters to describe Darwinian populations: H (fidelity of heredity), V (abundance of variation), S (dependence of fitness differences on intrinsic character), C (continuity), and α (reproductive competition). The aim was to say what distinguishes paradigm cases, the ones that give the Darwinian machine its importance, from both trivial cases and marginal ones—cases where the core Darwinian conditions are only approximated. In this chapter, three parameters have been used to describe reproduction in the case of collectives: B, G, and I. I now tie those two discussions together.

The different forms taken by reproduction have consequences for the features described with the first set of parameters. To examine these relations I will start paying more explicit attention to *levels* in the biological hierarchy. I understand talk of levels in a simple way, involving part/whole relations. Entities at level n are made up, at least in part, of entities at level $n - 1$. Organisms are made of cells. Social groups are made of organisms. Levels will be discussed more generally in the next chapter. In this section I focus just on one case, the relation between the evolution of integrated multicellular organisms like us, and the cells within us.

Humans form a Darwinian population—we vary, reproduce, and inherit various characteristics. But so do some of our parts, including cells. They too vary, reproduce, and pass on many of their characteristics in reproduction. In a collective of this kind, a threat is posed by "subversion" through independent evolution of the lower-level entities. If a cell arises that has a feature that makes it able to divide faster than others, and the feature is reliably passed on in reproduction, we expect that feature to proliferate, whether or not that feature does any good for the whole organism. So how do collectives like ourselves remain viable? Sometimes, of course, we do not. Cancer is one consequence of cell-level Darwinian processes (Frank 2007). But once we see the in-principle possibilities here—once we see ourselves as collectives with Darwinian parts—it can be surprising that we hold ourselves more-or-less together at all. Recent biology has been very interested in the mechanisms by which such subversion is prevented. In this section I will review some of these ideas, recast in the present framework.

Both organisms and the cells within organisms form Darwinian populations, but a number of features of complex multicellular organisms partially suppress the evolutionary activities of their cellular parts. I will refer to this as the partial "de-Darwinization" of lower-level entities by evolution at the higher level.

Bottlenecks are one such feature (Grosberg and Strathman 1998). This is the second theoretical role of B mentioned above. It can be initially surprising that large organisms start life so small—that the single-celled stage has been preserved. Smaller things tend to get eaten by larger ones, and there are often other advantages to being big. So rather than starting small and racing to *get* big, why not *start* bigger? One consequence of narrow bottlenecks is that they ensure an initial uniformity in the cells making up a single organism in the next generation. The result is a limited scene for evolutionary activity.

To say this is to treat the cells within a single organism as a Darwinian population, not to recognize a population comprising the cells within *all* humans (for example). I will return to this issue later. For now, the point is one about each population of cells that comprises a human organism, considered separately. Then a bottleneck constrains V (the abundance of variation) for that small population. The process of cell division starts with a common genotype. Variation will arise, both genetically and "epigenetically," but there is reduced scope for evolutionary competition.[7]

There is reduced scope, I said. But if an organism only has a bottleneck, the cell-level activity that *does* occur has definite consequences, not just for the lives of the organisms as individuals, but for evolution at the higher level. Suppose, to pick a simple case, that cells divide to form a large organism, but then mere chance determines which adult cell will become a propagule or spore that initiates a new organism. If the process is a matter of chance, then the spore-like cells will be representative of which cell types have done well, reproductively, during the lifetime of the organism up to that point. Then we expect cell lineages that increase their own representation within the multicellular organism to arise and spread—to spread not just within organisms during their lifetime, but also across generations of those organisms as well. A bottleneck, we saw, reduces the scope for evolution at the cell level within each individual, but does not make such evolution any less consequential. It might then seem that the new generation of multicellular organisms is being initiated at each step by successively better competitors in the within-organism context (Michod 1999).

This apparent problem may or may not be a serious one, depending on the biology of the organism, and there are other features of the situation that will be discussed later. But one point can be made now. Consider what happens if there is early sequestration of a germ line. Then however much somatic evolution there is, it becomes, in one sense, irrelevant. When a cell lineage competes successfully within the organism, this may affect how *much* the organism reproduces, but it will not affect the *composition* of the cells that initiate the next organism-level generation.

In my framework, this involves the role of the parameter S. When (for example) a lottery determines which cells will be propagules, those cells with intrinsic features that make them successful in the within-organism competition may be able to dominate in the longer term as well. When there is sequestration of a germ line, the features that make for within-organism success do not have this role. The cells that can become the precursors of a long lineage of cellular descendants are distinguished by their *location*, their relations to other parts of the organism.

---

[7] Earlier I left some vagueness in whether an absolute or relative measure of B was more important. In the clear cases we have both. But as John Matthewson pointed out, an absolute measure is more important in this de-Darwinizing role, while a relative measure is probably more important in the earlier role concerning the supply of variation in the evolution of collectives.

This is the argument that was sketched briefly in Chapter 3 when S was introduced. When an organism has a sequestered germ line there are still cells with high fitness and cells with low. But in a collective of that kind, the heritable intrinsic features of cells have limited long-term importance. A Darwinian process can still occur in the shorter term. When the immune system adapts to a new disease-causing invader we have one case of this; when a person gets cancer we have another. But this within-organism evolution has an end, and the only cells that can generate a long lineage after them are those in the germ line. Except in unusual cases, a cell cannot get itself into the germ line from outside. That does leave the possibility of competition *within* the special arena provided by a germ line, as will be discussed in Chapter 7.

So when there is high G at the organism level, there is reduced S at the level of cells. The presence of a sequestered germ line in a multicellular organism partially de-Darwinizes the cellular population below. Looking back to the spatial representation in Figure 3.1, we can imagine two populations moving simultaneously in the space. A population of collective entities would begin as a marginal case, with haphazard heredity—to the extent that reproduction can be defined at all—and other non-paradigmatic features. As the collectives become integrated and develop specialized reproductive machinery, they may move towards the paradigm region of the space. But the acquisition of a germ line implies changes to the status of another population, the population of cells. Cell division remains a process with reliable inheritance of variation, but many of the fitness differences between cells are now disconnected from intrinsic character: low S. They move, in at least that respect, away from the paradigm region. So both populations move through the Darwinian space of Chapter 3. And this is happening in part *because* of what the collectives, the new Darwinian population, are doing in the space describing reproduction (Figure 5.1).

Here I have looked at the consequences of bottlenecks and germ lines; this may or may not tell us about their origins. These questions will be discussed in more detail in the next chapter.

5.4. Marginal Cases and Reproduction

[*Persistence and metamorphosis; replacement and multiplication; species selection; ramets, sex, and origin explanations.*]

The picture I have developed holds that Darwinian change requires reproduction, but only in a permissive sense. Darwinian change can occur even when reproduction is poorly distinguished from other things. But paradigm Darwinian populations tend to have definite parent–offspring relations linking the individuals that comprise them. Marginal Darwinian populations often have modes of reproduction that are marginal themselves. A "marginal" case of reproduction is not one that *looks* strange given our everyday experience. To me, at least, the alternation of generations is strange, but there is nothing marginal about it. The marginal cases are those in which reproduction is unclear in a way that makes a Darwinian difference.

I will illustrate this by first looking at a limiting case. A number of people have noted that, from a formal point of view, asexual reproduction of one offspring plus death of the parent (or fission plus death of a daughter) seems not much different from continuation of the parent. Once we note that fact, we bump into a problem mentioned at the end of Chapter 2: why is reproduction needed for a Darwinian process at all? If selection changes a population by deleting some and retaining others, why isn't that enough?[8] I dealt with this earlier by treating selection without reproduction as either a part or a pale analogue of the Darwinian process, but not much was said about why that is. The point might have seemed to be mainly a verbal one. The issue is also liable to get tangled in debates in metaphysics: some philosophers argue that the persistence of any physical thing is a causal matter, in which earlier temporal stages cause later ones (Loux 2002). On this view, persistence itself might be seen as a kind of reproduction.

A reply can now be given that is neither verbal nor metaphysical. If persistence is analogous to a kind of asexual reproduction, it is a very marginal kind. In the simplest examples—as when an object persists from one day to another—there is no bottleneck in the process, and no other reorganization of the entity. A more interesting case to think about is one where a metamorphosis step breaks down and reconstitutes the individual's organization. This is similar to a bottleneck, and gives a non-arbitrary divide between "generations." In fact, many kinds of metamorphosis, especially in insects, include the death of a large majority of cells in the organism's body. Biologists wrestle with the distinction between reproduction and metamorphosis themselves (Bishop et al. 2006).

If these are cases of reproduction, however, they are cases where each individual has a maximum of one offspring. They do not include the possibility of *multiplication* (Maynard Smith 1988), but only *replacement*. Consequently, the only way there can be fitness differences is for the population to get smaller. Its evolutionary possibilities are very limited; selection cannot play a role in origin explanations, for example, in the ways discussed in Chapter 3. When "reproduction" does not include the possibility of multiplication, the result is at most a low-powered Darwinian process.

---

[8] Fagerstrom (1992) gives an analysis of fitness that views persistence and reproduction as equivalent, and Bouchard (forthcoming) gives an argument that persistence of lineages should *replace* reproduction in such an analysis. See also Darden and Cain (1989).

This extreme and simple case can be used to cast light on others. The differential persistence and proliferation of higher-level entities like species and clades (whole branches of the tree of life) has frequently been seen as Darwinian (Gould and Eldredge 1977, Williams 1992, Lloyd and Gould 1993, Gould 2002). Even aside from the strong causal hypotheses required by this idea, the entities in question sometimes do not look like they could engage in reproduction. In the case of "clade selection," Okasha (2003) has argued that there is a logical impediment to the idea itself, because a clade is supposed to include within it, by definition, *all* lineages descended from a given species. The only way for a "parent" clade to cease to exist is for all lineages descended from it to cease to exist, so a clade's children cannot outlive it, as a matter of logic.

Let us take "species selection" first. Species are very large collective entities, with little integration (though this may be controversial), and no germ line. Some ways in which new species appear do involve a bottleneck, however, and these are often seen as the most important kinds (Mayr 1963). A small number of individuals become isolated, and take a new evolutionary path as they multiply. So collections of species are unlikely to form paradigm Darwinian populations, but the idea of species selection is not especially far-fetched within the present framework. (This "founder-effect" speciation could be entered towards the $(\frac{1}{2}, 0, 0)$ region in Figure 5.1.) When we consider elements of the tree of life even larger than species, the idea of reproduction becomes more tenuous, however. I see the logical difficulties that Okasha cites as a symptom, rather than as the core, of the problem. There are probably ways of re-defining "clade" so that the idea of clade reproduction is at least coherent (Haber and Hamilton 2005), but it will be a very marginal kind of reproduction indeed. And clades might be differentially eliminated, but that is not enough for a significant Darwinian process.

To finish this section I will return to some problems discussed earlier, now that my treatment of reproduction is on the table. These are the problems with ramets, genets, sex, and identity.

Many biologists dealing with "modular" organisms, like corals and plants, treat the genet (genetic individual) as the fundamental evolutionary unit. Any extension of a genet through space counts as growth (Janzen 1977, Cook 1980). As noted earlier, this view has strange consequences when we extend it to single-celled organisms. Another view holds that whenever a life-cycle goes through the bottleneck of a single-cell stage, this marks a new generation, and hence a case of reproduction, regardless of the genetic relationships between the entities on each side of the bottleneck. This view is defended by Dawkins (1982a), citing Harper (1977).[9] The idea that *any* one-cell stage marks a case of reproduction has

---

[9] Though Dawkins cites him as his source, Harper's own view in this area seems more complicated. Harper says (1977: 27n) that any one-celled stage marks a case of reproduction, but

awkward relations to the phenomenon of metamorphosis, however, as was also discussed above. Sterelny and Griffiths (1999) use this as an argument against Dawkins. Many organisms exhibit narrowings, to various extents, at different stages in their life-cycle. In some parasites, which metamorphose trying to get into a host through a physical barrier, the narrowing goes down to a single cell. When we think about those cases, the genet-as-unit view may seem powerful again. Depending on the circumstances in which one wants to disperse, or the barriers one has to get through, one may break down more, or less, of one's body before rebuilding it.

Here is how those problems look in the light of this chapter. "Reproduction" encompasses a range of phenomena, including more evolutionarily significant ones and more marginal ones. B is one marker of this distinction, for the case of collectives. But reproduction without the possibility of multiplication is—in all cases, not just collectives—an evolutionarily weak or marginal kind. Cases where metamorphosis borders on reproduction, due to a bottleneck, are like that. If there is reproduction, it is a mere matter of replacement. In the case of collectives, G and I are also important; if only one pregnant buffalo makes it through the gap, that alone does not make herds into Darwinian individuals. The evolutionarily important sense of reproduction is not one constrained by rich criteria for "individuality," especially genetic criteria. And the very idea of a genet or clone as something bound together by genetic *identity* is, in most cases, an idealization.

These ideas about reproduction, bottlenecks, and sex are based not in intuition, but in an independently motivated account of Darwinism and Darwinian explanation. I will finish with an illustration of that fact. Consider a collection of plants producing new physiological individuals through apomixis, the asexual production of seeds. In Chapter 3 I gave a schematic description of how natural selection can figure in origin explanations. It does so by altering the array of genetic backgrounds against which new variants appear. Suppose we have two genotypes, one that is a single step away from a genetic combination that will produce a particular new phenotype, and one that is many steps away. Selection

he also says that it is the genet, the sum of *all* the genetically identical material produced from a zygote, that is the bearer of fitness (1985: 5; Harper and Bell 1979, 30). I am unsure how to square the two views, as I assume that whenever something reproduces it thereby has fitness. This explains why Harper is sometimes described as holding a bottleneck criterion of reproduction (Dawkins 1982a, and J. Wilson 1999) and sometimes as holding that reproduction requires the establishment of a new genotype (Vuorisalo and Tuomi 1986). Similarly, Jackson, whom I quoted in Chapter 4 claiming (with Coates) that genets are the "fundamental units" on which natural selection acts, says in a 1985 paper that he will treat reproduction as "any increase in the number of physically separate individuals by either clonal or aclonal means" (1985: 298). A glimpse into the psychology of biologists battling with these issues is provided on the same page, where Jackson speaks of the "nightmarish reexamination" of concepts necessitated by clonality.

can make the new phenotype more *likely* to appear by making the precursor combination more *common*. This increases the number of ways in which a single new mutation can result in the appearance of the new phenotype. When this sketch of the role of selection in origin explanations was given earlier, the example used did not specify whether reproduction was sexual or asexual, and phenomena of this kind are compatible with asexual reproduction. Sex, when it is present, has its own importance to origin explanations, as the fusion of gametes brings genetic material arising in different lineages together. But the role of selection in origin explanations outlined above does not require sex. Consequently, if what was given in Chapter 3 was a fair description of this role that natural selection has, then evolution by natural selection is compatible with asexual reproduction even for collectives, and even for collectives that *can* also have sex. The distinctive feature of the process described there, again, was that via the precursor type becoming more common, extra "slots" or "opportunities" were created in which a single mutation suffices for the new phenotype. This increase in the commonness of the precursor could be a matter of extra apomictic individuals being produced.

Above we imagined a precursor to a novel phenotype becoming more common via apomixis. We can also imagine, instead, the production of precursor ramets. Is that equivalent? In some ways, it is, but there is a crucial difference, and that difference is *B*, the bottleneck parameter. A single mutation can do more to produce a novel phenotype when it appears in an apomictic seed than it can in the meristem of a ramet. A ramet is not a fresh start to the developmental process in the wholesale way that a seed is. That is not to say that nothing significant can result from mutation on a ramet, as all drinkers of champagne will recall. We should raise our glasses to mutations in ramets. But we should raise them higher to mutations in seeds.

5.5. Summary of the First Five Chapters

Darwinian populations are collections of things that vary, reproduce at different rates, and inherit some of this variation. The basic features of these collections are startlingly routine—births, lives, and deaths, with variation and inheritance. But Darwin saw that this set-up, this arrangement of ordinary features, is an extraordinarily important element of the world. Darwin's description was empirical and concrete. The last century's work has included a series of moves towards abstraction, attempting to say what is essential about the Darwinian machine—which features are not dependent on the contingent particularities of life on earth. I continue that tradition, but do so with an eye to another feature of the Darwinian world view. Darwinian populations shade into marginal cases, and the paradigm Darwinian processes depend on ingredients that are themselves evolutionary products and must have come from something simpler. One aim of this book is to give an account of the Darwinian process that is designed to handle this blending-off into marginal cases, precursors, and not-quites.

This account of Darwinism yields a particular picture of the world. One of the world's constituents is a great range of Darwinian populations: paradigm cases and marginal ones, some clear and others obscure, some powerful and others suppressed. Some are visible and obvious, others invisible. Some are inside others. They tread through their Darwinian behaviors on a great range of different scales in space and time. Some evolve via reproduction of a wholesale and definite kind, others evolve by coopting the biological scaffolding that results. Populations evolve as a consequence of their Darwinian properties, but also change the basis for their further evolution, moving through the imagined spaces of evolutionary parameters. The tree of life is generated by Darwinian populations and what they do—the tree is a structure of lives linked by reproductive events. But reproduction is an evolutionary product, and appears as a different relationship at different places on the tree. Sometimes there is sex, a fresh start, and genetic novelty with every birth; sometimes the appearance of a new organism is imperfectly distinguished from continuation of the same thing. Some Darwinian individuals live inside others, in ways that make it unclear how to count and distinguish them. And sometimes the tree shape is lost due to fusions and hybridizations.

....

CHAPTER 6

LEVELS AND TRANSITIONS

6.1. Levels

[*Hierarchies; uniform application of Darwinian criteria; the replicator approach; exclusion of cases on causal grounds; gestalt-switching.*]

This chapter is about the idea that Darwinian processes occur at different "levels." We have encountered this theme several times in earlier chapters, but here it receives a more systematic treatment, with comparisons to other work on the topic.

The biological world is hierarchically organized. This is true in several senses, but one sense involves parts and wholes. Genes, roughly speaking, are parts of chromosomes, which are parts of cells. Cells are parts of multicellular organisms, which are parts of social groups and subpopulations within species. These, in turn, are parts of species themselves. Each of these entities has parts beside those just mentioned, but the ones mentioned (among others) are parts that can reproduce. At any level where reproduction is found, there is the possibility of a Darwinian process.

There can be reproducing entities that are not part of a standard hierarchy like the one above. These include tight symbiotic associations, like lichen. Algae and a fungus are the parts of a lichen, but the lichen is not a part of a "next entity up" in a standard hierarchy. One response to this is to recognize additional hierarchies (Eldredge 1985). Another is to rethink things like lichen. But the cleanness of hierarchical descriptions is not important here; more simply, the world contains reproducing entities at many different *scales*, that can at least potentially form Darwinian populations. As a consequence, some Darwinian individuals are physical parts of others, and have their evolutionary activities embedded within a special context—the evolutionary activities of another Darwinian population.

The approach taken to these issues here is a direct application of the framework in the preceding chapters. Suppose we wonder whether there is a Darwinian process at the level of cells, or of social groups. To answer this question, we should apply the usual Darwinian criteria to those entities. Do cells reproduce? Do they vary? Do some of the variations affect their rates of reproduction? And so on. This is a simple approach, and not a new one. In Chapter 2 my discussion of "classical" summaries of natural selection included Lewontin's 1970 formulation. That summary was given by Lewontin in a paper on "the units of selection," aimed explicitly at these issues. Lewontin's claim was that once we have isolated the essential features of a Darwinian process, and described these features in an abstract way, the resulting criteria can be applied to entities at many levels and scales. We apply the Darwinian criteria *afresh* at each level, leaving aside what we might have said about levels above or below.

The classical summaries of Darwinism have been pulled apart, put back together, modified and augmented in the last hundred or so pages, but that aspect of Lewontin's approach was right. Many others have disagreed, however, either explicitly or implicitly. So my aim in this chapter is to re-assert this simple handling of "levels" questions, and develop it further with the aid of ideas in the chapters above.

A Darwinian population is a collection of entities in which there is variation in character, the inheritance of some of those characteristics, and differences in how much individuals reproduce. These populations can be found at many levels. An old tree again provides an illustration. Consider a large oak, with many branches that originate deep in its history. The tree is a collection of cells, living and dead. The apical meristems, growing points on branches, are arenas of tiny and localized evolutionary events. A mutation arises via cell division, in one layer of a meristem. It may or may not become established in that layer. If it does, that branch diverges genetically from the others. In plants like the oak, between the level of the cell and that of the whole organism (or ramet) many biologists also recognize the level of the *module* (White 1979, Tuomi and Vuorisalo 1989a, Preston and Ackerly 2004). These are, roughly speaking, the units visible between branching events. They are also independent sites of sexual reproduction. Modules are born, live, and die: "a large mature tree with its myriad branches is really a population of modules with a distinct age structure" (Gill et al. 1995: 426). The tree's shape is itself a representation of a Darwinian process of differential reproduction and divergence at the module level. Above cells and modules there is evolution at the level of oak trees as wholes, reproducers of new trees. And there may be other levels beside these.

The tree, whose very shape reflects evolutionary processes within it, is a clear illustration. But the same picture applies more generally, including to ourselves. When we think this way, we employ the account of Darwinian processes in a uniform way at each level. So cases of "multi-level selection" are simply those where a system contains Darwinian populations at different levels, all evolving. It is significant, then, that much of the literature in this area has not applied a view of this kind. Sometimes the reason is the adoption of the replicator approach. This view holds that questions about "levels" and "units" in a Darwinian context are always ambiguous, as there are two roles that need to be filled in any evolutionary process. First, there must be entities at some level that act as replicators—entities that are faithfully copied. Second, there must be entities—perhaps the same, perhaps different—that act as "interactors" or "vehicles." These are the entities whose interaction with their environment leads to the differential copying of the replicators. There may be a hierarchy of such interactors, all with different effects on the copying occurring at the replicator level (Brandon 1988, Lloyd 1988, 2001).

The replicator approach was criticized in general terms in Chapter 2; replicators are not necessary. Consequently, there is no need to bifurcate questions about levels or units of selection into two, thinking that replicators must be present somewhere although the obvious entities in the process do not pass the replicator test.[1] Questions about the "unit" of selection are not ambiguous; the units in a selection process are just the entities that make up a Darwinian population at that level. It is always *possible* to ask a further question: what is the mechanism of inheritance? But that is an optional question about how the patterns of inheritance that give rise to a Darwinian process at a given level come about.

For other writers, there is a different reason why the simple Lewontin 1970 view has seemed inadequate. This has to do with the feature I described by saying that we apply the Darwinian criteria *afresh* at each level, leaving aside what we might have said about levels above or below. The objection often made is that an apparent Darwinian process can be a mere byproduct of a selection process at another level (Williams 1966, Sober and Lewontin 1982, Sober 1984, Lloyd 1988, Okasha 2006). The formal "shape" of a Darwinian process can appear at one level when the real process of natural selection is either above or below. Thus some prima facie Darwinian processes can turn out to be spurious, shadows cast by natural selection at another level.

For example, suppose (adapting an example due to Williams) that herds of deer differ in running speed. Fast herds do well in escaping predators, and slow herds do badly. Suppose also that successful herds also tend to give rise to new herds. But we might then learn that a "fast herd" is really no more than a herd of fast individuals, and a slow herd a herd of slow ones. Fast individuals reproduce more than slow ones, and fast individuals also happen to live together, producing fast herds. But there is no significant interaction or coordination between individuals in a herd with respect their running and escape behaviors. Then, many will say, natural selection for speed has not acted *on herds*, because the causally important properties are at the level of individuals. Herd-level speed is a mere byproduct of individual-level speed. The lives and deaths of individual deer have led to the

---

[1] "Levels" and "units" questions are also treated as synonymous here, though some have distinguished them (Brandon 1988). The same applies to questions about units of "selection" and of "evolution," distinguished by Maynard Smith (1988).

evolution of individual speed, and have generated various facts about herd speed as byproducts.

So a number of people have searched for some causal test—using relations between probabilities, regression coefficients, or other formal tools—that will distinguish the real cases from the illusory ones that should be analyzed as selection at a different level. The search has been frustrating.[2] It is easy to see some reasons why that might be so. Even in the case of a very clear higher-level selection process, there will be *some* explanation in lower-level terms for why things happen as they do. It is easy, when following this road, to rule too many cases out.

I will suggest a different way of looking at the problem, where we augment the simple Lewontin approach not with an extra causal test of the usual kind, but with some concepts used in the middle chapters of this book.

When people discuss "fast herd" cases, they do not usually say much about what herd-level reproduction is supposed to be—if, indeed, it is in the picture at all. This is partly, I suspect, because of the influence of a replicator/interactor picture, or something close to it. People suppose that the first problem is to work out whether herd-level properties are having an effect on reproductive differences *of some sort*; they do not necessarily have to be differences in *herd-level* reproduction. But that move is what sets us down the wrong road. If I believe that herd speed is a byproduct of selection on individual deer, then I am committed to the idea that there have been differences in the reproduction *of deer* that arose from differences in running speed, and that has led to the world being full of fast deer. I then add that when these deer form herds, we get a fast herd. If, on the other hand, I believe that herd speed has been the product of selection on herd-level properties, then I believe that there have been differences in the reproduction *of deer herds* that arose from differences in running speed. (I may believe both). If I favor this second view, though, I have to grapple with the idea of herd-level reproduction—with whether there is such a thing, and

---

[2] These proposals are lucidly reviewed in Okasha (2006). I will give an example of the problems faced that is especially illustrative here. Okasha locates two main contenders for criteria that will isolate the level at which selection is occurring. These are the "Price equation" approach (see Section A.1) and "contextual analysis." The Price approach has the apparent problem that it does not disqualify the fast-herd cases sketched above. As long as fast herds contain faster and hence fitter individuals, the Price approach recognizes selection on herds. A partial regression approach, contextual analysis, does disqualify these cases. But contextual analysis counts "soft selection" as higher-level selection. When there is soft selection there are no group-level reproductive differences at all. Both of these results seem completely unacceptable given the usual goals of the literature. Insofar as my view can be mapped onto this debate, I am closer to the Price approach, treating contextual analysis as a test for something important but different. It is a test for when group context or other population-level properties *make a difference to* a lower-level selection process, rather than a test for higher-level selection.

whether it is the kind of reproduction that can figure in a significant Darwinian process.

In most of the cases used to argue that apparent selection processes can be artifacts or byproducts of others, the illusory-looking case is also a *marginal* one, in the terms of the framework developed here. The problem is not (continuing with the same example) that speed of herd is too simply related to speed of individuals; the problem is that the herd is not the sort of thing that can figure in a significant Darwinian process at all. Individual deer, in contrast, are the sorts of things that can figure in paradigm Darwinian processes. These are reasons to favor the first of the two hypotheses above, the one that treats herd speed as a byproduct of the evolution of individual speed. Putting it another way, if there *was* reliable inheritance of variations in herd speed in a big collection of clearly reproducing herds, and it also turned out that the speed differences had a simple causal basis in individual leg muscle differences, that fact would not *prevent* the herds from exhibiting a Darwinian dynamic.

Putting it metaphorically, it would be possible to have one paradigmatic Darwinian population "sitting on top of" another one. There are reasons, discussed later in this chapter, why that might be an unlikely thing to find. But there is no need to augment a simple and direct handling of levels questions with an additional causal test. The cases that motivate those moves can be handled using the criteria discussed earlier in this book.

Following up this approach generates new puzzles, however. These arise especially when there are definite parent–offspring relations at two different levels. Then the reproduction of some lower-level entity partly comprises reproduction at the higher level, and is partly distinct from it. The case of us and our cells is an example. Suppose we say there is a Darwinian population at the cell level as well as one at the level of organisms like us. Are we saying there is one Darwinian population of cells within me, and another within you—a collection of separate Darwinian populations—or is the idea that we can think of all the human cells as comprising one big Darwinian population? The answer is that we can see the situation in both ways; both kinds of collections qualify as Darwinian populations. In the case of cells, biologists usually focus on the separate populations within each organism. (This is discussed as "somatic" or "developmental" selection.) But both kinds of analysis are possible, with the two kinds of populations having different evolutionary parameters.

The cells within a single human form a small population, tightly bound by their ecology. This population has low genetic variation. (It varies more in the marks established by "epigenetic" inheritance mechanisms—Jablonka and Lamb 1995—but for now I will focus just on genetic variation.) The totality of human cells alive at some time also forms a Darwinian population, but also an unusual one, for different reasons. This collection of cells is genetically diverse, but is packaged into discrete groups (organisms like you and me), which are genetically very similar internally. Most genetic variation is across these groups, not within them, but most interaction is within them.

This way of thinking can be illustrated by drawing an analogy with social groups. Suppose that a range of groups of humans is initiated from pairs of similar individuals, resulting in a collection of sealed communities that are internally homogeneous but very different from each other. Such an image lends itself to several kinds of evolutionary analysis. We can first consider evolutionary processes within a single sealed community. There will be some initial genetic differences, and new genetic variation will also arise. Darwinian processes will ensue. Secondly, there may be an evolutionary process in which whole communities are reproducing entities. Each community, if it does not go extinct, might send out pairs of individuals to form others. And thirdly, we can recognize a human-level, as opposed to community-level, process *across* the ensemble of these communities, even though they are largely sealed. This would be especially clear if the formation of new communities involved individuals from two different communities forming the pair that initiates a new one—sex at the community level, as well as the organism level. But an evolving human population could be recognized even if the communities reproduced asexually.

Humans remain Darwinian individuals in such circumstances; they are still born, live, and die, varying and inheriting, even though their activities are tightly "packaged" into collectives. That is so even if there is a "germ line" at the community level, so that only some lineages of individuals within a community can give rise to the colonists that initiate new communities. Descriptions of the three different kinds could also be combined. The processes within each social group combine as parts to make up the overall evolution of human individuals. The group-level processes are visible when we "zoom out" from various lower-level ones.

We can think the same way about us and our cells. There are Darwinian processes within humans, Darwinian processes in which whole humans form the population, and Darwinian processes at the level of human cells in general. The three kinds of description fit together, in principle, but the processes have their own distinctive Darwinian features when considered separately. The small cell-level populations have little genetic variation but lots of cell-level interaction; the larger cell-level population has lots of genetic variation but this variation is packaged into units that mostly interact internally. That makes the total population of human cells very different from a typical population of bacterial cells, or protozoan cells. In those cases there is much less packaging into genetically similar clumps, so genetically different individuals interact more than genetically different human cells do.

What we are engaged in here is a kind of "gestalt-switching," analyzing Darwinian populations at different scales and with different boundaries (Kerr and Godfrey-Smith 2002a). As we do this, parameters describing variation, interaction, and inheritance change.

6.2. Cooperation and Altruism

[*Altruism as an evolutionary problem; kin selection, reciprocity, and group selection; attempts to assimilate all to group selection; abstractions of kin selection; correlated interaction.*]

The biological characteristics most often at the center of debates over levels of selection are cooperation and altruism. It initially seems that familiar Darwinian processes at the level of individual organisms must always favor selfish, exploitative behaviors over cooperative ones. Certainly it is hard to see how natural selection would favor "altruistic" traits, which involve one individual giving away its resources to others, or taking risks on behalf of others. But these behaviors are often seen in nature (see Dugatkin 2002 for an empirical review). The relations between "cooperative" and "altruistic" traits will be discussed below; for now I will use the term "altruism" in a broad way, as is common, to cover the entire category.

There is a standard menu of evolutionary mechanisms that can, in principle, explain how altruistic behaviors can survive in an evolutionary context. *Kin selection* (or "inclusive fitness") hypotheses are based on the idea that an individual can increase the representation of its *own* genes in future generations by helping *others* reproduce—provided those others are likely to carry similar genes to the actor (Hamilton 1964). Mechanisms of *reciprocity* involve one individual donating resources to another, in the expectation that the recipient will reciprocate (Trivers 1971, Axelrod and Hamilton 1981). *Group selection* hypotheses suppose that although selfish individuals do better than altruistic ones within groups, in some cases altruism can flourish because groups with many altruists are more productive overall, or less likely to go extinct, than predominantly selfish ones (D. Wilson 1980).

Categorizations of options such as this one are continually being contested, especially with respect to which mechanisms are primary and which might be special cases of others (Sachs et al. 2004, Lehman and Keller 2006). For some years it was common to argue that group selection was only likely to be important when groups were formed by biologically related individuals, so significant cases of group selection will be just special cases of kin selection. More recently, some have argued that group selection provides the fundamental mechanism, and the other options rely implicitly upon it. Below I will defend a particular way of looking at the options. I will arrive at this view by discussing some other attempts at assimilation and unification, however, beginning with views that emphasize the role of groups.

Here is a standard model. We assume biological entities at two levels. The lower-level individuals reproduce asexually, with high fidelity and discrete generations. Their life-cycle includes the regular formation and dissolution of groups. Specifically, in the initial stages of life the lower-level individuals collect into groups of a fixed size, via some rule. Interactions within these groups affect the fitness of the members, and the reproduction stage in the life-cycle includes the dissolution of the groups, forming a new "pool" of juvenile individuals. Then the cycle begins anew (D. Wilson 1975, Wade 1978). This is sometimes called a "trait-group" model, and its structure is illustrated in Figure 6.1.

The lower-level individuals are found in two types, **A** and **B**. (The "**B**" here is emboldened to distinguish it from the *B* parameter of Chapter 5). The model assumes that being of the **A** type carries a direct cost, and being *around* the **A** type is associated with benefits. It is as if the **A** type "donates" some fitness to everyone in its group. So, for all individuals, the more **A** individuals there are in their group, the fitter the individual is. And for any given group context, it is better to be a **B** (not paying the cost) than to be an **A**.

More details are given in the Appendix. The main result is that the **A** type can do better than **B**, overall, even though *within* any group context, **B** does better than **A**. For **A** to do better overall, however, the groups must be formed in a

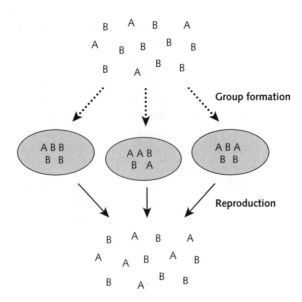

Figure 6.1: A life-cycle with ephemeral group structure.

particular way: there must be a tendency for like to interact with like, at a rate greater than chance. A individuals must tend to find themselves in groups with other A individuals, and B with B. Then when the A individuals donate fitness to others, the benefits go mainly to others of the A type. Depending on the details, the A type may then survive and proliferate despite the fact that whenever As are in a group with Bs, the Bs successfully exploit them.

For some, this schema is *the* key to the evolution of altruism. What look like other mechanisms really involve the same process in slightly different guise: "The theories that have been celebrated as alternatives to group selection are nothing of the sort. They are different ways of viewing evolution in multigroup populations" (Sober and Wilson 1998: 57). The groups may persist for longer times (with rounds of reproduction within them), or for shorter times; they may be transitory interactions of the kind modeled with game theory. The individuals might be kin, or not. Either way, it is the relation between within-group and between-group advantages that determines whether altruism will survive.

In fact, however, groups of any kind are optional (Maynard Smith 1976, Godfrey-Smith 2008). Figure 6.2 pictures a life-cycle involving a different kind of population structure. Juveniles settle not into groups, but into a lattice, where each interacts with its immediate neighbors (only its north–south–east–west neighbors, we can suppose). As before, the reproduction step dissolves the lattice. Again there is an altruist type and a selfish one. Altruists pay a direct cost, and

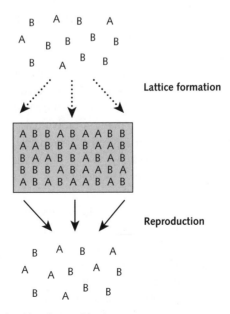

Figure 6.2: A life-cycle with ephemeral lattice structure.

any individual receives benefits from each of its A-type neighbors, if it has any. So the fittest thing to be is a B type surrounded by A; the least fit is an A surrounded by B. The A type can survive and proliferate in such a regime, provided that the rule by which individuals settle onto the lattice is one that tends to bring A individuals into contact with other A individuals at a rate sufficiently higher than chance. The Appendix discusses more details; the analysis is similar to the one that applies to Figure 6.1, except that groups do not exist. Groups are replaced, in a sense, by neighborhoods. But there are as many neighborhoods as there are individuals. Neighborhoods, unlike groups, cannot be seen as collectives competing at a higher level.

The comparison of the cases in Figures 6.1 and 6.2 does two things. First, it shows us that group structure, even of a fleeting kind, is not essential to the evolution of altruism. Second, it suggests what *is* essential. The common element in the two models, when altruism does survive, is the presence of *correlated interaction*. In the case in Figure 6.1, A individuals tend to find themselves in groups with other As. In the Figure 6.2 case, As find themselves encountering other As as neighbors. Either way, the result is that the benefits donated by the A type tend to fall on other A individuals. A non-random division of the population into groups is one way of achieving this situation; another is the non-random distribution of individuals on a lattice or network.

Correlated interaction is the key to the evolution of altruism. Group structure, preferential interactions among kin, and reciprocity are all ways of achieving the right kinds of correlation. This idea is a theme in the literature that has frequently been seen, but sometimes then obscured. I will discuss it in more detail after I have looked at kin selection. First, I will use this discussion of groups to say a bit more on the topics of the previous section.

Let us look back at the scenario in Figure 6.1. Clearly there is reproduction by the lower-level individuals, along with variation and heredity. These individuals also form groups. And in at least one sense, these groups may do well or do badly in relation to each other; groups with many A individuals are more productive than groups with fewer. But for a group to be "productive" here means that it gives rise to many lower-level individuals. And change in such a model is usually measured as change in the frequencies of different types of lower-level individuals (A versus B).

Groups are clearly *important* in this scenario, but there is debate over whether this counts as group *selection*—selection on groups as units. On one view, the model sketched Figure 6.1 is one in which selection occurs at two levels. B does better within groups, but A can prevail because of group-level advantage. If the arguments earlier in this book are right, the description given of the case so far does not support that interpretation. "Group selection" requires a Darwinian population *of groups*; groups must vary, reproduce, and inherit features from other groups. So far, all we have is reproduction and inheritance by individual organisms, affected by the groups in which they are found.

Maybe half the writers in this area agree with that last claim, and half do not. The result has been an occasionally tense debate about proper terminology (Sober and Wilson 1998, Maynard Smith 1987, 1998). Some peace has resulted from making a distinction between two senses of phrases like "group selection" and "multi-level selection" (Damuth and Heisler 1988). In "MLS1" (multi-level selection of type 1), groups play a role in an evolutionary process, but a "group fitness" is just the sum of the fitnesses of the individuals within the group, and change is measured by tracking change in the frequencies of lower-level individuals. In "MLS2," the fitness of a group is measured in terms of its production of *offspring groups*, and change is measured by tracking change in the ensemble of groups.

The model in Figure 6.1 is usually thought of as a case of MLS1. However, it is *possible* to describe this scenario in terms of group-level reproduction. We might say that group X is a parent of group Y if one of X's constituent individuals has offspring that become part of group Y. This is a possible description but a strained one. A group may have as many parents as it has constituent individuals. Other oddities arise as well (Okasha 2003a). This, I suggest, is another problem that can be put into a new light via the framework in this book. Some formal and experimental models of "group selection" feature definite forms of group-level reproduction and some include only marginal ones. The Figure 6.1 case is a marginal one; it is *barely* describable in terms of group-level reproduction. New groups are formed by aggregation, with no bottleneck or germ line. Other models imagine groups budding or dividing, like plants or cells. And a few are organized with an eye to collectives that engage in very clear collective reproduction, like bee colonies and ourselves (Wade 1978, Keller 1999). A good way to make progress on the cases that debate continues to swirl around might be to start with broad and inclusive sense of "group-level reproduction" but then sort the clearer cases from the marginal ones.[3]

I now leave the topic of groups. The next item—and often the main one—on the list of ways of thinking about the evolution of altruism is kin selection. The original idea, developed primarily by William Hamilton (1964), was that an

---

[3] Another illustration is provided by the work of Wade and Griesemer (1998, Griesemer and Wade 2003). This is an unusually thorough investigation of group-level heritability and its consequences. An experimental study was done in a way featuring a clear parent–offspring relation at the group level—new demes were established from a small number of old demes. But many of the broader conclusions drawn concern the feasibility of the "shifting balance" evolutionary process of Wright (1932), in which *migration* between *existing* demes is the main mechanism by which group fitness is "exported." That is a process in which there is only very questionable group-level reproduction.

individual can further the proliferation of its own genes by helping its relatives to reproduce. The genes that proliferate will then include those that are the basis for the helping behaviors themselves. "Hamilton's rule" summarizes the idea, for one class of cases, with the idea that an altruistic act will be favored via a particular kind of behavioral interaction if $rb > c$, where c is the cost to the actor, b is the benefit to the recipient, and r is the "coefficient of relatedness" between them.

That is the original idea, but the last forty years have seen a series of reformulations and re-derivations of Hamilton's principle. These reformulations have had a definite direction, broadening and abstracting it.[4] The result is that "relatedness" in its normal sense is now optional. The core principle can be summarized by saying that an altruistic act will be favored if the actor's behavior leads to benefits falling on individuals who tend to bear the heritable basis for that same action. This will be discussed in detail in the Appendix, but I will follow the trail a little further here, as it motivates some of my conclusions below.

One reformulation of Hamilton's rule is due to David Queller (1985). Queller gave a derivation in which the familiar cost (c) and benefit (b) of altruism still appear, but relatedness is replaced by a measure of covariance, or correlation. What is measured is the covariance in the population between the *phenotype* of the *actors* (the individuals producing, or not producing, the key behavior) and the *genotype* of those the behavior *affects*. Informally, altruism survives if the benefits of the acts fall disproportionately on those who are likely to pass the behavior on. Queller noted briefly that the same principle will cover cases of reciprocity, the third category on the standard list of mechanisms, and Fletcher and Zwick (2006) have used this version of Hamilton's rule to analyze classical models of reciprocity (such as the Iterated Prisoner's Dilemma) and have shown that the principle is so broad that it covers cooperation across *species*. In the light of this follow-up work, Hamilton's rule only has a contingent connection to relatedness.

These points can now be put into an overall picture. A number of writers have said, often quickly, that the familiar mechanisms behind the evolution of altruism can be seen as different ways of achieving correlation between the traits or behaviors exhibited in a population—a tendency for like to accompany like (Hamilton 1975, Eshel and Cavalli-Sforza 1982, Michod and Sanderson 1985, Sober 1992, Skyrms 1994). The idea has sometimes been conflated with others, and accompanied by claims at tension with it. But the convergent message of the recent literature is along exactly those lines. Breaking it into two: altruistic individuals can have higher overall fitness within a time period or generation if behaviors within the population are correlated; altruism can survive and

---

[4] These include Hamilton (1975), Michod and Hamilton (1980), Queller (1985), Grafen (1985), Frank (1998), Fletcher and Zwick (2006), and others.

proliferate across generations if the benefits of altruism fall on those with a tendency to pass the behavior on.

One path to this picture starts by thinking about groups, notes the importance of "clumping" of types in group formation, and then bumps into the fact that correlation of the right kind can arise without groups being present at all. Another path starts from the idea that kin selection is the most empirically well-supported mechanism in this area, but notes that once we give a formal description of what is essential to kin selection, relatedness is optional and the heart of the matter is an abstract measure of correlation. Here is another way to put it, using yet another path. With or without "altruism," a behavior flourishes in an evolutionary context if *somehow* it differentially generates fitness benefits for those with a tendency to transmit the behavior to their offspring. One way for this to happen is for the actor (producer of the behavior) to do things that increase the actor's *own* fitness—assuming that the actor is likely to pass the behavior on. Another way is for the actor to produce behaviors that help other individuals who are likely to repay the favor, aiding one's own "direct" fitness by a more circuitous route. A third is for the actor to help others who may not repay the favor, but who have a tendency to pass on the behavior when *they* reproduce. These others may be close relatives of the actor, but need not be. The second of these two possibilities, the one involving circuitous effects on an individual's own reproduction, is more naturally called "cooperation." The third, helping another reproduce, is more naturally called "altruism." The underlying principle is the same.

It is also clear, when we follow this third path, that there need not be some particular gene associated with a particular altruistic behavior, a gene that is reliably copied and is a kind of "long-term beneficiary" of the evolutionary process. Suppose that altruism in some population appeared as a quantitative trait, like height. Everyone has different degrees of altruism, but the more altruistic tend to bestow benefits on the more altruistic. The covariance expressed in modern versions of Hamilton's rule applies to a quantitative trait with a mixed basis for inheritance. We can understand the persistence of altruistic behavior without positing an underlying and selfish long-term beneficiary.

6.3. Transitions

[*New Darwinian populations from old; a schematic transition; subversion of higher levels versus suppression of lower; fraternal and egalitarian transitions; B and G as de-Darwinizers; sex and other complications.*]

This chapter so far has looked at the relations between evolutionary processes at different levels while assuming that both the higher- and lower-level entities are present. But, of course, the entities that populate these various levels and scales arise *via* Darwinian processes. This has been a prominent theme in recent work, exemplified especially by Maynard Smith and Szathmáry's *The Major Transitions in Evolution* (1995), and before them by Leo Buss's *The Evolution of Individuality* (1987).

The term "major transitions" in evolution is used in broader and narrower senses (Sterelny and Calcott, forthcoming). In the broader sense, the transitions are evolutionary events of any kind that greatly alter the character of evolution downstream. The narrower sense is what I have in mind here. These are what Michod (1999) calls "transitions in individuality"—transitions that involve the origin of new kinds of biological individual. Two crucial examples are the evolution of the eukaryotic cell and the evolution of multicellularity.

In my terms, a transition in individuality involves the appearance of a new kind of Darwinian population. That is what gives us the important sense of "individual" in this context—the appearance of new entities that can enter into Darwinian processes in their own right. In many cases, including the two mentioned above, this involves changes to the status of collective entities. They begin as combinations or associations of different reproducers, with only a marginal mode of reproduction at the collective level. The initial association might be between two prokaryotes, one who has freshly engulfed the other, or between two single-celled eukaryotes who remain stuck together rather than separating after cell division. But in time, these collectives come to engage in a definite form of reproduction in their own right, and the collectives may come to form a paradigm Darwinian population at the new level. That is half the story, however. The other half is a set of changes to the evolutionary status of the lower-level entities that went into the collective. These typically move *away* from paradigm status. Their independent evolutionary activities are curtailed, constrained, or suppressed by what is happening at the higher level—a partial "de-Darwinizing" of the lower-level entities. In three central cases—the eukaryotic cell, the multicellular organism, and eusocial insects—this process took different paths.

I will approach these issues by first working through an idealized example—a schematic transition, with some of the flavor of several real-life ones. Then I discuss complications.

Imagine we have a Darwinian population of free-living entities that has all the features of the paradigms. For simplicity, assume its constituent individuals reproduce asexually and use ordinary genetic mechanisms of inheritance. These individuals then—for whatever reason—come to live in groups of some sort: colonies, herds, or clumps. Individual-level reproduction continues, and new groups also regularly appear, being formed from fragments of one or more old groups. As discussed earlier, it may be *possible* to recognize reproduction at the level of the collectives in such a situation, but as an equivocal or marginal case. Then, however, we imagine an increase in cohesion and integration in these collections, with extensive networks of cooperation and mutual dependence. Group living becomes less and less optional.

Collectives of this kind face problems of subversion. Networks of cooperation create possibilities for exploitation by free-riding mutants. A collective of entities which retain the capacity to individually reproduce risks devolving into a Darwinian "tragedy of the commons" (Hardin 1969, Kerr et al. 2006). But we can now introduce a series of factors which alter the situation.

One is a bottleneck. If a new collective arises from a single lower-level individual, then the result, at least initially, will be a collective that is genetically uniform. The consequences of a bottleneck can also be described in terms used in the previous section. If we think of the lower-level entities as forming a large population packaged into groups, a bottleneck is one way of creating correlated interaction in that population.

Suppose, then, that collectives arise through a bottleneck, but chance determines which lower-level entities will become the propagules that initiate a new collective; there is a sort of germinal lottery. Then the propagules will be representative of which lower-level individuals have done well, reproductively, during the lifetime of the collective up to that point. Things are different again, however, if there is a germ line. Suppose some of the collectives acquire this feature; early in each collective's life, one lineage of lower-level individuals becomes specialized for propagule formation, and the others become unable to take on this role. Now if one type competes successfully within the collective this may affect how *much* the collective reproduces, but will not affect the *composition* of the propagules that initiate the next collective generation. A typical initiator of a new collective is no longer a descendant of an individual who has done well in a within-collective competition, but a descendant of an individual in the germ line of a successful collective. Here we imagined a sequence in which the collectives acquire a series of anti-subversion devices. But this is *also* a sequence in which the collectives have successively acquired several features, discussed in Chapter 5, of a definite mode of reproduction at their own level.

This transition was described in an abstract way but has the flavor, of course, of a transition from single-celled to multicellular organisms like ourselves. It has this flavor because of the particular ways that subversion problems were handled. There are other ways things might have gone. In the scenario above, the consequences of lower-level competition were allayed by sequestering a germ line. Another way of dealing with this problem would be for one member of the collective to prevent reproduction altogether by other individuals within the collective. The initiator remains intact and lays an enormous number of eggs generating the entire collective, like a queen bee.

A third path would be one in which the collectives have no bottleneck, being formed from larger fragments, but one member of each new collective has partial control over the lives and activities of the others, especially with respect to their reproduction. A good way for one member to achieve this control over the others might be to swallow them, as one cell did to another about 2 billion years ago (Lane 2001).

The transition to the eukaryotic cell is often considered different in kind from the other two compared to it here. In Queller's terms (1997), the transition to multicellularity is a "fraternal" transition and the evolution of the eukaryote is an "egalitarian" one. In an egalitarian case, two very different kinds of entities come together, bringing different capacities that work well in combination. The problem they face is the possibility of competitive exploitation. In the fraternal cases, the entities are initially similar. The "hurdle" such a transition faces is whether there is initially anything to be gained by it. The benefit may come from advantages of scale, rather than the fusion of different and complementary capacities. This can be put more explicitly in terms due to Calcott (2008). For a transition to occur there must somehow be both the *generation of benefit* and the *alignment of reproductive interests*. Both Queller and Calcott think there is something like a trade-off between the two. Subversion in the fraternal cases can be allayed by close kinship within the collective, and the problem is generation of benefit. This assumes, however, some mechanism that generates highly correlated interaction in the fraternal cases. Otherwise subversion is a problem to be solved.

So there are several ways of "de-Darwinizing" lower-level entities in a transition, with three possibilities exemplified by the cases of eukaryotic cells, multicellular organisms, and eusocial insects. In each of these, an initial collective has come to engage in definite high-level reproduction, and this has involved the curtailing of independent evolution at the lower level.

In the rest of this chapter I will discuss the case of multicellularity in more detail. First I will follow up the logic of the idealized case above a bit further, and then look at some further empirical features of the case.[5]

As discussed earlier, there are two ways of looking at lower-level entities once they have become packaged into tight collectives but can still reproduce. One way is to see each collective as the arena of a separate and short-lived Darwinian process. The other is to look across the total population of lower-level entities. This is a case where gestalt-switching can be usefully done.

If we start treating each collective separately, things look like this. Collectives formed through bottlenecks have low variation (V), when compared to collectives

---

[5] This discussion has been influenced by Buss (1987) and Michod and Roze (2001). For a different perspective on these issues, emphasizing concord rather than conflict between levels, see Otto and Hastings (1998).

formed in other ways. So there is little fuel for evolutionary change. (To say this is to simplify things, because there can be heritable variation arising rapidly via the "epigenetic" cell-level systems of inheritance.) The second perspective involves tracking change in a total lower-level population spread across the collectives. Now we assume a larger and more variable population, and the role of a bottleneck at the collective level is to "package" the variation in a particular way.

In Chapter 3 I used the parameter α to represent the degree of competitive interaction within a population. We can think of any population that is divided into sealed collectives as one in which there is lots of interaction within groups, and much less between them. When the sealed groups are formed via a bottleneck, the individuals that interact intensively are genetically similar. This gives us a different way of looking at situations of highly correlated interaction in a population: the relations between individuals who differ in evolutionarily relevant ways feature weak interaction; the strong interactions are between those who are similar.

So perhaps the treatment of α should be adjusted to take into account this relation between α and V. Paradigm Darwinian populations feature not just genuine competitive interaction *somewhere* in the population, but interaction between individuals who do vary. Cells organized into internally homogeneous collectives are less Darwinian than cells in heterogeneous collectives or no collectives at all.

I now turn to the consequences of a germ/soma divide at the collective level. A germ line does not have special consequences when we think about each collective as the site of a separate Darwinian arena. The germ line is just where some of the lower-level entities live (often quietly). The more significant relationship is between the population of collectives and the total lower-level population distributed across the collectives. Then there is a connection between G at the collective level and S at the lower level.

S, again, is the extent to which differences in realized fitness depend on intrinsic differences between members of the population. The link between G and S was discussed in the previous chapter for the case of organisms and their constituent cells. When there is a germ line in an organism like us, there are still cells or cell types with high fitness, and cells with low. This is true in both short-term and longer-term senses. In the short term, a cell may acquire intrinsic features that enable it to give rise to many daughter cells. But the only cells that can generate a long lineage after them are those in the germ line. And these cells are distinguished largely by their location, their relations to other parts of the organism. This point applies to the lower-level elements of all collective reproducers. When there is high G at the collective level, there is reduced S at the lower level. As high S is associated with paradigm cases of Darwinism, the presence of a germ line in a collective partially de-Darwinizes the population below.

This relation is made complicated by the fact that fitness differences between collectives also affect S at the lower level. If one collective has very low fitness in comparison with others, suffering an early death, then *all* the constituents of that collective have their fitness pulled down. And if collectives are internally homogeneous ones, then the lower-level entities that die will have intrinsic features in common. Here we again run into the fact that makes the situation complex. In a case like this, to some extent the facts of lower-level reproduction *comprise* the facts of collective reproduction, and to some extent they are distinct from them.

Here I have discussed the de-Darwinization of the parts of complex collectives. It is possible for this sort of process to go so far that the system loses the distinctive features of a population altogether. This theme will be discussed in more detail in Chapter 8, but I will introduce the idea now. Not every configuration of matter, not even every collection of distinct parts, is best seen as a *population*. Once things are too tightly bound into a network, with highly asymmetric roles that derive from their place in that structure, populational concepts lose their grip. A large organic molecule, for example, is not naturally seen as a population of atoms. This is not just because the atoms do not reproduce, as not all populations are Darwinian. A riot is a populational phenomenon, even if no one is reproducing.

In organisms like us, cells reside in organized networks and have lost a significant part of their autonomy. We have taken some steps away from being populations of cells at all. We have not taken *many* steps though; the cells within us maintain their own boundaries and retain the capacity to reproduce. They retain crucial Darwinian features in their own right. They have lost the autonomy enjoyed by their distant ancestors, but they have not become like mitochondria—they have not given up essential parts of their reproductive machinery. (Red blood cells, which lose their nuclei as they mature, are an exception.) Our cells are not—or not yet—"post-populational." Less tightly bound organisms, like corals and trees, retain more thoroughly population-like relations between their parts.

This theme is also related to one discussed by Dan McShea, who argues that as collective entities become more complex and integrated, their component individuals tend to become simpler in structure and behavior, often losing parts over evolutionary time (Anderson and McShea 2001, McShea 2002). This may happen for various reasons, and different kinds of parts may be lost. From the present point of view, there is a special status to the loss of reproductive machinery.

The last part of this section will bring the discussion of multicellularity above into contact with further empirical details and the history of life. The first complication comes from the fact that before the evolution of complex multicellularity, many single-celled organisms were already making extensive use of sex—engaging in complicated sequences of mitosis and meiosis, splitting and recombining their genetic material. The evolution of multicellularity in plants, animals, and fungi occurred on a cell-level backbone that included "ancient haploid/diploid cycles," to use Maynard Smith and Szathmáry's phrase.

With collectives made up of entities that reproduce only asexually, there can easily be a bottleneck without a germ line, and vice versa. But sex itself makes bottlenecks likely (Grosberg and Strathman 1998, Wolpert and Szathmáry 2002). Part of this is related to the problem of internal competition; even if all the cells in a particular parent contributing gametes to a large propagule were themselves very genetically similar, the gametes would differ, owing to the many separate events of recombination that scramble genes and chromosomes during sex. So sex magnifies the problem of internal conflict. There may also be other problems with the possibility of a large sexual propagule. Grosberg and Strathman suggest mechanistic problems with the coordination of cells during the earliest stages of development. Wolpert and Szathmáry claim that it would be hard for such organisms to have a coherent developmental program. As it has turned out, bottlenecks without germ lines are common (Buss 1987), and germ lines without narrow bottlenecks are not (as far as I know) found at all.

In Chapter 5 I emphasized the effects of bottlenecks on the evolutionary potential in a population, by making special kinds of variation available. As was noted then, it is a further claim that bottlenecks evolved *because* they do this. Above I discussed the role of bottlenecks in the prevention of subversion. That might explain the prevalence of bottlenecks, but it might be a byproduct too. As far as origins go, in a sense the bottleneck (one-cell stage) came first and the bottle followed. As far as retention is concerned, bottlenecks might be retained in many populations because of the requirements of sex, complex development, or something else again.

I will also discuss complications with the role of G, and the germ/soma divide. Plants do not have germ lines, in the sense of cell lineages "sequestered" for the function of producing sex cells. I categorized plants in Figure 5.1 as having an intermediate value of G, as some cells go down a path that normally prevents them from acting in the reproduction of new plants. But without a germ line, one might wonder how large plants have remained viable in the face of subversion problems. If germ lines are needed for large animals like us to keep afloat, how do plants do without them?

Part of the answer may be that subversive cells are not nearly as destructive in plants (Buss 1987, Klekowski 1998). Cancer-like growths are common in trees, in fact, and they seem to do nothing like the harm they do to animals like us. Part of the explanation for this, in turn, seems to be the fact that plant cells cannot move around the organism, but are fixed in position due to their rigid cell walls. Plant tumors do not spread nearly as dangerously as animals ones do. Another part of the explanation may be the fact that plants are less interdependent in their organization than animals like us; malfunction in one place does not so frequently lead to disaster for the whole. So if we ask who *does* need a germ line, given these theoretical ideas, it will be creatures with a high degree of organization and interdependence, large size and a long life, and cells that can move around within the organism.

As populations evolve they change the features that are bases for their further evolution. These features include the nature of reproduction. Some of these features, such as bottlenecks and germ lines, have consequences for the downstream evolutionary possibilities in a population. Changes to the nature of reproduction will not usually happen *because* of those consequences, but once they occur, for whatever reason, new evolutionary doors are opened.

CHAPTER 7

THE GENE'S EYE VIEW

7.1. Genes and Darwinian Populations

[*Genes as parts of organisms; genetic accounting; selection of scaffolded reproducers; transposons, homing endonucleases, meiotic drive.*]

For many biologists and philosophers there has been an elephant in the room throughout the last few chapters. A very tiny elephant, but a watchful and omnipresent one. The elephant is the gene, and the possibility of dealing with many or all of these problems from a "gene's eye view." This is not just a matter of making more use of genetics; it involves taking the perspective of individual genes as evolutionary units.

The "gene's eye view" is associated with a family of ideas, some more empirical and some less so. First, there is the idea that in *some* cases, perhaps very special ones, genes are the units of selection rather than organisms or other entities (Burt and Trivers 2006). A second, and quite different claim, is that *all* cases of biological evolution (or almost all) can be *represented* that way. The gene's eye view gives us one available description that coexists with other descriptions of the same cases. To this may be added the claim that a few phenomena *only* have a gene-level description (Sterelny and Kitcher 1988). Then there is the strongest version of the view, seen in Dawkins (1976), which holds that it is a mistake to describe most cases as selection on anything else. Natural selection is a contest between replicators, and genes are almost the only replicators—certainly organisms and groups do not usually qualify.

There has been a great variety of philosophical criticism of these ideas, mostly directed at the second two views listed above. (See Okasha 2006 and Lloyd 2006 for reviews.) But against this there can be found not just philosophical defenses but a growing list of empirical cases where the gene's eye view seems to help—a list of phenomena that we have got a *handle* on via this way of thinking. Some of these cases may be accepted by critics of the gene's eye view, as a short list of special phenomena that can be accommodated in a multi-level picture of evolution. But then we have to work out why the gene-level description is appropriate in those cases and not in others. For defenders of a gene's eye view, the "special" cases are just those where the general nature of gene action is seen in a pure form.

The treatment here proceeds, again, by application of the framework defended in earlier chapters. This makes sense of the appropriateness of a gene-level description in some special cases, explains the artificiality of descriptions in purely genic terms in other cases, and explains how the two are related. Here is an outline of the main points. To treat genes as "units of selection" in some situation is to treat them as making up a Darwinian population. Roughly speaking, genes are scaffolded reproducers; the copying of DNA is a kind of reproduction. In some respects, genes are like cells: they are small parts of organisms which, because they can reproduce, make up lower-level Darwinian populations. Genes are also central to inheritance in both cells and whole organisms. As a consequence, a lot of evolutionary change at those other levels can be *tracked* in genetic terms. Much of that description has a special status that is easy to misread. It often appears to be a description in which a Darwinian pattern of explanation is applied to genes, when in fact the Darwinian description is being applied to *organisms* that are described and categorized in terms of their *genetic properties*. In addition, however, there are some phenomena in which genes enter into distinctive Darwinian processes of their own. Most of these cases are the products of the details of eukaryotic sexual machinery. From an evolutionary point of view, in fact, the very existence of genes as units is dependent on this modern sexual machinery. The question of why that machinery exists then looms large.

The status of genes as Darwinian individuals is shown to be quite different from the other cases discussed in this book. By some strict but reasonable standards, genes as evolutionary units do not exist at all. By more relaxed standards, they do. Evolutionary explanations at the genic level require the relaxed standards. And "selfish gene" cases do not function as models or exemplars, in which the overall nature of evolution is seen in a pure form.

To begin, let us look at the place of genes in the biological hierarchy. Cells are parts of organisms. Chromosomes are parts of cells. And genes are parts of chromosomes. This last claim might be met with some unease, for reasons discussed below. But initially at least, I will treat genes as small parts of living things.

In the preceding chapters, when low-level parts of organisms were discussed I usually chose cells. Cells give us a partial model of how to treat genes, but the two are different in several ways. First, genes are scaffolded reproducers, unlike cells which are simple reproducers. DNA is replicated as part of the process of cell division, via the larger machinery of the cell. In a sense, cells too cannot reproduce "on their own," as environmental conditions need to be suitable. But in the case of genes, the required environmental conditions are *very* specific—they need to contain almost all of the machinery of genetic reproduction. In addition, the role of genes is made complex by sex and meiosis, which create cells with new combinations of genes derived from separate sources. In the previous chapter I noted that in the case of cells, biologists usually focus on evolution in the separate populations *within* each organism, rather than the total population of (say) human cells. In the case of genes, in contrast, people usually think in terms of a population extending *across* organisms—evolution in the total pool of human genes. But in principle, both kinds of analysis can be done in both cases.[1]

This is also an area where the literature has evolved special "mixed" ways of talking that do not apply Darwinian concepts in a straightforward manner. I will give two illustrations. First, within a simple and direct treatment of genes as potential Darwinian individuals, *all* the gene-sized bits of DNA are potentially competing with each other. It is often said, in contrast, that a gene competes only with its alleles at a given locus. This organizes much talk of genetic competition, but it is also known not to be true. Various phenomena associated with "transposons"—genes which move around within the genome—show the falsity of the claim. Those phenomena are, and are accepted as, Darwinian. The "only compete within a locus" rule is not one that genes themselves respect. Genetic competition occurs across loci as well as within them; these are just different environments which a genetic element can occupy.

Second, a lot of talk about the evolutionary role of genes is really talk about organisms, genetically characterized. This can be seen in the accounting. In standard models, a diploid organism will be described as *AA, Aa,* or *aa* at a particular locus. It might be said that the *A* allele is selected over the *a* allele, for example, when the fitness of the *AA* combination is higher than that of *Aa,* and that of *Aa* is higher than that of *aa.* This will lead to an increase in frequency of *A* over *a.* This familiar description is more unusual than it looks. If there are more *AA* organisms than *aa* organisms, in the sense above, that does not imply there are more physical copies of the *A* allele than of the *a* allele in the population. The *aa* organisms might contain many more cells than the *AA* organisms, and hence there may be more physical copies of *a.* In the standard accounting, each diploid organism is counted as equivalent, and each contributes two units to the calculation of the genetic composition of the population at that locus.

So there are different possible ways of counting genes in a population. Call the "standard count" the one that counts each organism as equivalent regardless of the number of cells and hence gene copies. The alternative one, which treats all gene copies on a par, can be called the "simple count." The special features of the standard ways of accounting show up also in the treatment of loci. When a

---

[1] A cellular population within an organism like us is asexual. The "total" population of human cells, in contrast, shows asexual cell division, reductive cell division producing gametes, and fusion of gametes. So it is a complicated system—but so are many protist populations.

gene jumps to another locus, the standard count does not treat this as an extra copy of the old gene. It is a new allele at that locus. Within mainstream models, talk about how selection leads to the proliferation of a gene (and so on) is usually a mixture. Some of it is directly aimed at tracking the spread of copies of a gene. But much of it is really talk of the natural selection of organisms, genetically characterized.[2]

It is no accident that organisms are being genetically described and categorized. Genetic properties do not merely give a handy label. Gene action is often causally responsible for one organism reproducing more than another, and genes are also central to the patterns of inheritance in organism-level populations. So genetic properties are pivotal to the evolutionary role of organisms. But this is a situation in which a Darwinian population is being recognized at the level of organisms—they are the things whose reproduction we are tracking and explaining—and that Darwinian population is being affected by the genetic properties of its individuals. This is one reason why genes, as discussed in evolutionary biology, sometimes seem partially abstract, less than fully material: talk of genes is not being used to refer to physical particulars made of DNA, but as a way of talking about sharable properties of organisms.

Let's think about what taking a *purely* gene-level view would involve. Shifts of this general kind were discussed in Chapter 6, for the case of organisms and their constituent cells. In the cell case, the shift in perspective is mainly a matter of "zoom," or how closely we look. Organism-level reproduction mostly *is* cell-level reproduction, suitably organized. The total goings-on at the cell level comprise most of what goes on in organisms. There we have a relationship between simple reproducers and collectives. That is not true in the case of genes. In the genetic case, an analogy that can be used is one of "staining" the organism-level population, as when using a microscope. Suppose we could stain all the DNA in the world, in a way that makes the rest of each organism invisible. We then also zoom in. We will see a great collection of facts about genetic reproduction, variation, and inheritance. We will also see packets of genetic material grouped into various kinds of interacting clumps. Suppose we are looking at humans. Then there will be many small packets containing a fixed number of tangled strands. These packets are internally diverse except that particular pairs of strands within them are similar. These (cell-level) packets are collected into (organism-level) clumps that are mostly very internally similar across packets. Variation exists mostly across, not within, the organism-sized clumps. We also find that most

---

[2] Arguments that the gene's eye view is merely a matter of "bookkeeping" were originally developed by Wimsatt (1980) and Gould (2002). I add that the bookkeeping being done is not a simple and direct form of genetic accounting, but more a counting of organisms guided by their genetic properties.

of the copies of any stretch of DNA are, because somatic, dead-ends and the producers only of short lineages. A few give rise to long lineages, and they are distinguished from others in their clump largely by location.

When we look at a case of evolution this way, most of the machinery by which genes reproduce, proliferate, and have their effects on the world is missing—is not visible under the stain. A proponent of the gene's eye view does not deny the importance of that machinery, but adds it back into the picture as *background*, as a context to gene action, and an arena in which genes compete. Why might we have reason to look at a case like this? There is good reason to do so in some cases. These are the ones classified as "selfish genetic elements" (Burt and Trivers 2006). Talk of "selfishness" will be discussed below, but the key feature of these cases is that the crucial step in their working is that there is a gene-level reproductive difference that does not go *via* a difference made to organism-level reproduction.

This list of such cases is long and fascinating. I will discuss three examples (drawing on Burt and Trivers's review) which illustrate different forms of the phenomenon. Transposons, mentioned earlier, are genetic elements that move to different places within the genome. If the old copy is retained as new ones are added (which happens in some but not all cases), then the genetic element increases in frequency over others. The description of transposons puts pressure on familiar ways of talking about genetic competition, as it does not involve competition "for representation at a locus." But if mitochondria can compete reproductively within a cell, so can stretches of nuclear DNA. There are various mechanisms by which this reproductive advantage can be gained. One example was given in Chapter 4, in my discussion of "formal reproduction." A LINE transposon codes for an mRNA molecule which is translated to produce a couple of proteins that bind to the mRNA and reverse-transcribe the RNA back into the cell's genome in a new location. So there are now two copies of that element in the genome where before there was one.

This a case where a genetic element proliferates *within* a cell and *across* loci. There are also cases of proliferation within a cell and within a locus. "Homing endonuclease" genes exploit the cell's machinery for DNA repair. When a chromosome breaks in a diploid organism, the cell uses the other matching or "homologous" chromosome as a template to repair it. This is because as well as joining the break, some DNA often must be replaced around the two sides of the gap. If the homologous intact chromosome differs from the broken one (if the cell is a heterozygote at that locus), then this process of repair creates a new copy of the DNA sequence of the unbroken chromosome. Homing endonuclease genes take advantage of this fact. They code for an enzyme that cuts DNA at a specific site, called a "recognition sequence." The DNA that codes for the cutter is also inserted into the middle of the recognition sequence itself. This disrupts the recognition sequence, so the cutter does not cut itself. But in a heterozygote cell where one chromosome contains the cutter gene and the other does not, the cutter will break the other chromosome and thereby induce the cell to copy it into the other chromosome in the process of repair.

A third example is "meiotic drive." These genetic elements are diverse, but the general pattern is like this. A chromosome that can "drive" contains a "killer" element and a "resistant" element. In a heterozygote cell that is about to form haploid sex cells, there will be one chromosome with the "driving" complex and a homologous chromosome without it. Each sex cell produced will contain just one of the pair. The killer acts at some point during meiosis to sabotage the newly produced sex cell that contains the other chromosome. The driving chromosome has a "resistant" element at the place in the genome that the killer targets. This prevents the driving chromosome from destroying the sex cell that contains itself. This case is different from the other two because now the gene-level advantage does often go via a contribution to a difference in cell-level fitness. Whole cells are being sabotaged by the driving complex, those with particular genetic properties. In the cases of transposons and homing endonucleases, the process that generates a reproductive difference between genes takes place within a single cell.

So in the cases of transposons and homing endonucleases there is a gene-level reproductive difference that does not go via a contribution to cell-level or organism-level reproductive differences. In order for the genetic element to spread, the normal machinery of cell and organism reproduction must then enter the picture. But it makes sense to treat that machinery as mere background, because the crucial gene-level advantage was gained by processes within a cell. In the meiotic drive case, the gene-level advantage goes via a contribution to cell-level reproductive differences, but not via a contribution to organism-level reproduction. At least, that is true when the driving mechanism is acting alone, and often it does not. In many cases, an individual with two copies of a driving gene dies or is sterile. Even regardless of this, because of the role of cell-level fitness meiotic drive is a less pure case of gene-level selection than the others.

A biologist says: "A gene appears, which does X ... and it will proliferate." What this usually means is that an organism appears, with a new genetic property. The organism will reproduce successfully as a consequence. (The gene, given its context, makes for better camouflage, better disease resistance, a more impressive song.) The result is more organisms with that genetic feature. What the biologist sometimes means, instead, is that an organism with that genetic property will help *other* organisms with the same genetic property to reproduce. This is a case of the kind discussed in Chapter 6. If you help a brother, or a worker bee helps a queen reproduce, one organism contributes indirectly to the proliferation of organisms with its genetic features. Sexual reproduction leads to complicated cross-cutting patterns of genetic similarity in populations. The result is often a need for head-spinningly fine-grained genetic accounting (Queller and Strassman 2002).

Regardless, the Darwinian population in question is made up of reproducing individuals, with cells and genes as some of the individuals' parts. And what the biologist above *very* occasionally means is that the gene does something within individual cells to give it an advantage over other genetic material *in that cell*, or gives that cell an advantage over other cells within or produced by the same organism.

So in some cases it does make sense to focus on the activities of DNA per se, backgrounding most of the machinery of cell-level and/or organism-level reproduction. But those cases are unusual ones; to the extent that we organisms are products of gene action, we are not complex joint products of *that kind* of gene action. The picture, instead, is like this. To have evolution at all there must be some kind of reproduction, and evolution often gives rise to very sophisticated kinds. Once certain kinds of reproductive machinery are in place there is scope and space created for various additional Darwinian possibilities, via highly scaffolded, sometimes parasitic, reproduction. Given the presence of elaborate sexual reproductive machinery, we would *expect* some of this to arise. But these phenomena do not represent the general pattern of Darwinian evolution any more than parasites represent the general pattern of living activity.

Some of this picture can be seen in the Burt and Trivers survey I have drawn on in this section (2006: 25). Selfish genetic elements reliably arise, and tend to sweep through populations. But they often create conditions that undermine themselves; "selfish genetic elements almost invariably set in place forces that cause their own deterioration."

7.2. The Evolution of Genes

[*Genes as evolutionary units; dependence on crossing-over; team-shuffling analogies; evolution of recombination and the origin of genes.*]

This chapter has worked so far within a particular picture of what genes are like. Genes are treated as small stretches of DNA that are mostly preserved intact across generations while constantly entering into new combinations through sexual reproduction. This view is often described using analogies: genes are like cards that are repeatedly shuffled; genes are like rowers who are mixed into new teams (Dawkins 1976). Much of the time, that picture is accurate enough. But it involves an idealization, an imposed simplifying picture. In some contexts the idealization becomes misleading. A close look at where and how this picture breaks down leads to more conclusions regarding the status of genes as Darwinian individuals and units of selection.

I will introduce the main point immediately, and then approach it from several angles. When genes are recognized as units in an evolutionary context, a stretch of

DNA is said to count as a gene not only because of its effects on the organism, but because of how it is passed on. Genes are taken to have a degree of independence and persistence in this process. Thus a particular gene copy can, it is said, give rise to a definite lineage of descendants even though a chromosome cannot.

The fact that chromosomes cannot do this while genes can is a consequence of *crossing-over*: the exchange of genetic material between homologous chromosomes during meiosis. As advocates of the gene's eye view note, the "size" of a gene for the purposes of evolutionary explanation depends on the rate of crossing-over in the population. The way people commonly talk has it that the facts about crossing-over make chromosomes temporary, but leave genes as persisting and definite units. I will argue that this is not so.

This has consequences for evolutionary questions. In the most straightforward cases, a Darwinian population is made up of a set of definite countable things. Talk of genes as entities subject to natural selection relies on different and looser standards. My argument will not be that talk of gene-level reproduction and fitness makes no sense at all—the argument is not intended to contradict the previous section. But genes are not nearly as straightforward examples of Darwinian individuals as they look. In some ways they are marginal cases. It is not that there are no natural units at all in the genetic domain. Chromosomes and nucleotides are bounded natural units; we know where one ends and another ends. But the unit between these, the gene, is more dubious. In an evolutionary context it is more accurate to talk of *genetic material*, which comes in smaller and larger chunks, all of which may be passed on and which have various causal roles.

I will now go through these ideas in more detail. We can start with bacteria. How many genes are there in a typical bacterium? The standard answer is a few thousand (for example, four thousand in *E. coli*). This figure is basically a count of *cistrons*, genetic elements responsible for production of a single protein. The reality of these units gives us the length of each unit (a thousand nucleotides or so) and the rough location of boundaries between them. Then it seems that in a local population of a million bacteria there may be a few billion gene copies present (American billions, that is). We can count the bacteria, and we said there were a few thousand genes in each one.

But as many evolutionists will be quick to interject, this is not really the right count in an evolutionary context. Assume, for simplicity, that these bacteria do not engage in plasmid exchange, and the reproduction and spread of genetic material occurs only by simple cell division. Then the entire bacterial genome functions as an evolutionary unit; there is no basis for seeing it as a collection of distinct replicating things.

We now move to the case of humans, who are diploid and sexual. How many genes are there per individual? A standard figure is about 25,000. We could then go through, as above, a calculation for the number of gene copies in a local human population. Each cell contains two sets of the 25,000. We multiply by a few trillion for the number of cells in a human, and then multiply by the number of humans in the population. But that 25,000 figure is, again, not a figure that is directly applicable in the evolutionary context. It is again a count of cistron-like units. This count is more complicated and problematic than it was in the bacterial case, due to the elaborate organization of eukaryotic genomes, but let us accept it as close-enough to right—certainly it is not a completely arbitrary number.[3]

In the bacterial case, several thousand of these things were combined in a single genetic reproducing entity. How many such things are there in the human case? Chromosomes, at least, are easy to count. Suppose for a moment that no crossing-over existed in the human population. Then each human diploid cell would contain 46 units with a genuine evolutionary role; chromosomes would be high-fidelity scaffolded reproducers, reshuffled into new combinations during sex and gradually diverging along their own asexual lineages due to mutation. But in humans there is crossing-over, and chromosomes are not passed on intact. Advocates of the gene's eye view argue that this forces us to recognize smaller genetic units as replicating entities.

Suppose we tried to follow this logic in a pure way, recognizing genetic units simply on the basis of the facts about crossing-over. We then run into the problem that crossing-over does not respect the boundaries between cistrons, or any similar boundaries. Crossing-over breaks and recombines genetic material, and the only boundary marking discrete units that cannot be broken is the individual nucleotide. Crossing-over does not break chromosomes entirely at random; some regions do not break, and other regions are "hot spots" where it occurs more often. Most crossing-over events will hit non-coding regions, just because they take up most of the genome. But crossing-over does not reshuffle discrete genetic units in a way that respects functional boundaries between them. So the facts about crossing-over determine the *length* of a stretch of DNA that is likely to persist for a given period of time, but they do not determine the division of a chromosome into definite segments of that length. The lengths can start and finish anywhere. They are like stretches of time, not card-like units that are picked up and rearranged. Crossing-over may give us "units" in a units-of-measurement sense (the *centimorgan*, in fact), but not in the building-blocks sense.

---

[3] In the arguments in this section I will, for simplicity, often not make use of the fact that many "genes" with a known evolutionary role are not cistron-like at all, but are regulatory elements which have, given a specific context, definite phenotypic effects (Moss 2003). These phenomena strengthen the argument. For the ever-increasing complexities involved in counting genes as our knowledge of the organization of genomes grows, see Griffiths and Neumann-Held (1999), and Griffiths and Stotz (2006).

Summing up this set of arguments: the analogy with teams of rowers is a misleading one. Rowers are organized countable units that remain individually intact as they are shuffled into new combinations, and they have reasonably consistent causal roles within their teams. No genetic element in a population like ours has that combination of properties.[4]

I will also approach the point from a slightly different direction. Advocates of the gene's eye view say, as noted above, that the size of a gene in an evolutionary context depends on the particular rate of crossing-over (Williams 1966, Dawkins 1982a). As the rate of crossing-over gets higher, the number of genes presumably gets higher—or at least, it gets higher until there is a kind of collapse. Suppose crossing-over occurred every couple of nucleotides in each meiotic event; there is an almost complete re-shuffling of the sequences on the two chromosomes. Then, I take it, genes as evolutionary units would not exist at all; there would be no genetic element between nucleotide and chromosome that was copied as a unit.

The average rate of crossing-over in humans is roughly two times per pair of homologous chromosomes, per meiotic event. (The number varies across chromosomes of different sizes.) How do we feed that number back in to calculate the number of evolutionary genes in humans? At this point, defenders of the gene's eye view say that the size of an evolutionary gene depends not only on the crossing-over rate, but also on the strength of selection. Williams said in his classic 1966 book that a gene is any stretch of DNA subject to a "selection bias equal to several or many times its rate of endogenous change" (p. 25). Endogenous change includes both mutation and crossing-over. But we then encounter the fact that "selection differentials" change as the environment, the overall genetic composition, and the behaviors found in the population change. Genes will fade in and fade out of existence—though without changing their intrinsic physical features—as various selection differentials get larger and smaller.

The question "how many evolutionary gene tokens in humans?" is turning out to be unanswerable—not because we do not know the facts well enough, but because there is no definite number to learn. There is a fairly definite number of human beings, human cells, human chromosomes, and human DNA nucleotides. There is also a rougher number of human cistron tokens. But there is not a definite number of evolutionary genes, and the only sketch of a calculation that has been offered would yield a number that has an obvious element of arbitrariness and would also change as selection pressures change.

On one hand, there is an agreed-on set of facts about meiosis, selection, and the structure of genomes. On the other hand, there is a standard way of talking about genes, as units that are passed on intact while doing things that affect their

---

[4] Except perhaps for Y chromosomes, or at least the main non-recombining part of them.

rate of replication. I am arguing that the relation between the descriptive habits and the facts that provide the grounding for the habits is less straightforward than often supposed. When looking at genes in a close-up and empirical way, people discussing gene-level selection often use language that acknowledges these facts. Burt and Trivers, in their extensive 2006 review, do not talk of "selfish genes" but instead of "selfish genetic elements," a phrasing that steers the reader away from the idea that these are discrete units of the sort posited in classical genetics. I hear their term "element" not as suggesting something element*al*, but as referring to any *piece of genetic material* that has, given the local context, an ability to causally affect "its" reproductive rate in ways other than by fostering organism-level reproduction.

Dawkins himself argues that the sorts of facts raised here simply do not matter, as there is a harmless "elasticity" in the concepts of a replicator and evolutionary gene (1982a: 90). I agree that these facts do not matter much if one's point of view is sufficiently pragmatic. We can pick any piece of DNA found in some organisms in a population and note, given that specific context, its ability to causally affect its reproductive rate. When the context or our interests change, that bit of DNA will no longer be a salient-looking unit. But these considerations do matter if the aim is to give an account of the real entities that undergo the kind of change Darwin described.[5]

Within a more general discussion of genetics, Sterelny and Griffiths (1999) say that the word "gene" has become a "floating label" for any reasonably small stretch of DNA whose role is significant in the circumstances of a particular discussion. Whether this is generally true or not, it does apply to the case of evolutionary genetics, where the stretch of DNA that makes an evolutionary difference can be very unlike a classical gene (Moss 2003). At this point a defender of genic selection might argue that *all* the terms used to pick out alleged Darwinian individuals, including "organism," "cell," and "group," are floaters in this sense. Chapter 4 showed some buoyancy in the case of "organism," I accept, and looser collectives raise problems. But the problems with genes are more acute than the uncertainties found in the cases of organisms and cells; no one has argued that organisms fade into and out of existence as selection pressures change.

At this point it is useful to make a comparison with the status of populations themselves. There is some freedom, I argued, in recognizing the boundaries of

---

[5] Dawkins also addresses one of these issues directly. He says that crossing-over within a cistron will not usually break up a gene; only if the break is between two polymorphic sites will the old structure be lost. The rest of the time, it will be broken and then put back together (1982a: 90). This reply conflates *copying* with the mere fact of *reappearence* of a structure. A replicator is not faithfully copied if it is broken in half and then, owing to the luck of the gene pool, the same sequence is restored.

a population. Two evolutionary factors discussed earlier (sex and competition) affect how well "glued" a collection of individuals is into a natural unit. But as discussed in the previous chapter, it would be possible to recognize several small evolving populations and "stitch them together" to get a picture of evolutionary change in the larger collection. The situation with genes is different. It is not one where we first recognize a set of definite individual entities and then encounter some flexibility in how we collect them into groups for analysis. Here there is no clear inventory of the entities themselves.

So far the arguments in this section have been based on present-day facts about genes and chromosomes. To finish the section I will look at these issues from a more historical point of view. One part of the argument above can be summarized by saying that genes are only evolutionary units—to the extent that they are at all—as a consequence of crossing-over. That makes vivid the question: what is the evolutionary origin of crossing-over?

Crossing-over is one of three main kinds of exchange of genetic material. Bacteria engage in a kind of sex, "conjugation", which is severed from reproduction, by giving and receiving small packets of genetic material that may be integrated into a bacterial chromosome or carried around separately. (They also pick up stray bits of DNA that may be floating around, or gain them from viruses.) We can imagine a schematic history in which bacterial sex is initially the only kind of genetic exchange between organisms, until the appearance of the eukaryotic cell. In eukaryotes, the circular bacterial chromosome is replaced by some larger number of linear ones. And at some point, eukaryotes began to engage in cycles of haploidy and diploidy, requiring the successive doubling and halving of genetic material. At this point I will move the story forward within the framework of one particular hypothesis about the selection pressures that then took hold. This hypothesis is chosen provisionally, and because it throws some relationships into particularly sharp relief, not because I have special reasons to think it is superior to others. The hypothesis, developed by Haig and Grafen (1991), focuses on the role of intra-cell conflict.

Imagine a situation where an organism is cycling between haploid and diploid stages. A meiosis-like process forms haploid cells from diploid ones at a particular point, and it distributes half the diploid set of chromosomes into each haploid cell in the way familiar from ordinary meiosis, but with no crossing-over. This, Haig and Grafen argue, creates rich opportunities for destructive conflict. If a "killer" chromosome arose that could sabotage the haploid cell that the killer did not end up in, it would spread through the population. However, typical "driving" chromosomes, of the kind described in the previous section of this chapter, have two components, a destructive element and also an element that prevents the killer from destroying itself. In the driving complexes that are found these are tightly linked, and that is a clue to the hypothesis. If there was no crossing-over at all in a situation with a haploid–diploid cycle of this kind, the resources of entire chromosomes could be turned to the evolution of devious driving mechanisms. A killer and its protection could be far apart on the chromosome but still reliably passed on as a unit, and they could have subsidiary devices working with them on the chromosome as well. Crossing-over prevents such large weaponed complexes, such "terminator chromosomes," from being possible, as it breaks up genetic associations on chromosomes except for those that are very physically close (or protected by an inversion). The complex and technical part of the Haig and Grafen hypothesis is showing that these facts will actually select for genetic elements that foster crossing-over. Here I will assume their hypothesis gives a *bona fide* possible mechanism for the evolution of crossing-over, whether or not it is the actual mechanism.

If this was how evolution went, the resulting picture would be like this. Genes, roughly speaking, are late-comers. They are products of complex evolutionary measures taken by cells to suppress what would otherwise be carnage at the chromosomal level. Before the advent of haploid–diploid cycles in eukaryotes, genes as evolutionary units did not exist. I said "roughly speaking" because of the complicating role of genetic exchange at earlier stages in evolution. A more exact way to put it is like this. Gene-like units only have an evolutionary role as a consequence of *some* process of shuffling, so that small genetic elements can be passed on independently from others. If all we have is the reproduction of whole bacterial genomes, or whole eukaryotic genomes for that matter, then we may have identifiable cistrons (and various regulatory elements) but we do not have genes as evolutionary units. Bacterial conjugation is one kind of shuffling; segregation of chromosomes is another; crossing-over is another again. And the evolution of crossing-over is what set small genetic elements free as evolutionary players. This is the "evolutionary transition" that gave us genes. Not all transitions make big things out of small ones; the evolution of crossing-over created new small things out of bigger ones.

People often think of gene-like things as early arrivals in the history of life. In some "RNA world" scenarios, they are the earliest of *all* arrivals. If so, they were then largely *lost* as evolutionary players, for at least a billion years. Here I have in mind the long interval between the origin of the bacterial cell and the origin of something like eukaryotic sex. That event gave small genetic elements a role on the evolutionary stage once again. The trouble they periodically cause for the organisms that contain them is part of the "cost of sex" for organisms (though a very different kind of cost from the twofold cost that arises from the presence of males).

Even within the strong empirical assumptions made to tell this story, there are qualifications. Bacterial conjugation and related phenomena involve pre-eukaryotic shuffling, and may have had an extensive evolutionary role

(Woese 2002). In the previous section I also allowed that a simple increase in copy number of a genetic element within a cell, as with transposons, does have a Darwinian character. That could, in principle, happen without shuffling. But sex, which makes the fragmentation of genomes routine, set small genetic elements free in a new way, and the in-principle possibility here is also one that is important: an evolutionary transition that unleashes small things which were previously only aspects or variables characterizing a whole.

7.3. Agents, Interests, and Darwinian Paranoia

There are a great many important characters ... that are in the nature of collective attributes, all possessing the common quality of contributing to the welfare and survival of the group as such, and when necessary subordinating the interests of the individual. One of these is the reproductive rate. (Wynne Edwards 1962: 19).

Four thousand million years on, what was to be the fate of the ancient replicators? They did not die out, for they are past masters of the survival arts. But do not look for them floating loose in the sea; they gave up that cavalier freedom long ago. Now they swarm in huge colonies, safe inside gigantic lumbering robots, sealed off from the outside world, communicating with it by tortuous indirect routes, manipulating it by remote control. They are in you and in me; they created us, body and mind; and their preservation is the ultimate rationale for our existence. They have come a long way, those replicators. Now they go by the name of genes, and we are their survival machines. (Dawkins 1976: 19–20).

Treating genes as evolutionary units is often allied with a version of the "agential" approach to evolution, the approach that understands evolutionary processes as contests between agents with goals and strategies. The idea that genes are units of selection is often expressed by saying that genes are the units of evolutionary *self-interest*—the true contestants battling on evolutionary timescales, the agents who organic adaptations are "for."

Overtly agential description of evolution is part of a larger family, or a graded series, of metaphorically loaded usages. At the extreme end we have talk of strategies and cabals. These shade into less tendentious talk of welfare and goals, and those shade into talk of costs and benefits understood directly in terms of components of fitness—chance of survival, number of matings, and so on. It can be unclear where metaphor ends and literal usage begins. Talk of this kind can also have several different intended roles. It may be seen as a metaphorical expression of a deep truth (as in Dawkins 1976), or merely as a practical tool for thinking about some complex matters in a simple way (Haig 1997).

A good way to approach the status of this talk is to put it into a context provided by parts of recent psychology and anthropology, work which also casts new light on the history of ideas (Medin and Atran 1999, Griffiths 2002). Some of this work was outlined in Chapter 1. It argues that when dealing with the living world, people naturally make use of a particular package of conceptual tools.

These include an essentialist causal model of organisms, habits of teleological thinking, and a willingness to explain events in terms of agents and agendas. These habits do not operate the same way in all domains. The biological world triggers it more than some others, though Aristotle's science shows the potential for very systematic application of some elements of the package, and in the face of complexity and frustration it can be applied more broadly.

Descriptions of evolutionary processes often feature a mix of populational concepts with teleological and agential forms of description.[6] Shifts in views about how evolution works will then be accompanied by shifts in how agential talk is applied. This phenomenon can be illustrated by looking at some famous debates about evolutionary processes in the middle to late twentieth century, summarized in the quotes at the start of this section.

A number of evolutionary writers in the middle of the last century freely invoked selection at higher levels than the individual organism, especially in explaining cooperation and various kinds of restraint. Eventually there was a reaction against this thinking, spearheaded by Hamilton, Williams, and Maynard Smith. They argued that commonly invoked mechanisms of high-level selection would not in fact be evolutionarily efficacious, as lower-level evolution would lead to subversion of cooperative groups even if such groups were, in some sense, better adapted than non-cooperative ones. The attention to genetic models of such processes led to the development of the gene's eye view itself. Towards the end of the century there was a revival of explanations in terms of multi-level selection, but in a more rigorous form, as in the models discussed in Section 6.2.

This process was accompanied by successive shifts in use of the language of agency and benefit. Such language often has a significant communicative role. When a student is told that the gene is the ultimate unit of evolutionary self-interest, for example, he or she is supposed to hear that as gesturing towards one family of evolutionary mechanisms—which can be more precisely described in other terms—and away from another. In the case of descriptions from a genic point of view, however, these formulations developed an unusual power and role. They became more than a shorthand, being used not just to summarize complicated ideas but to shape foundational descriptions of evolution. An example of this was discussed in Chapter 2. There I discussed a passage by Dawkins in which an agential picture of evolution was used to argue for a requirement that any process of natural selection contain long-term persisting entities of some sort. (You cannot get evolution by selecting between things

---

[6] Influence of the older habits shows up in psychological work on the understanding of evolutionary ideas in students, even those who have had extensive instruction (Lombrozo et al. 2006, Shtulman 2006).

when there is only one copy of each.) I argued that this observation shows the limitations of agential descriptions of evolution, not a requirement of long-term persisting entities. This is truly a case of the metaphorical tail wagging the scientific dog.[7]

How did the gene's eye view acquire such apparent power as a foundational description? I conjecture that this is because the gene's eye view of evolution is a special kind of agential narrative.

Two explanatory schemata can be distinguished, within the general agent-positing category, which have a special psychological potency. The first is a *paternalist* schema. Here we posit a large, benevolent agent, who intends that all is ultimately for the best. This category includes various gods, includes the Hegelian "World Spirit" in philosophy, and includes stronger forms of the "Gaia" hypothesis, according to which the whole earth is a living organism. The second schema is a *paranoid* one. Now we posit a hidden collection of agents pursuing agendas that cross-cut or oppose our interests. Examples include demonic possession narratives, the sub-personal creatures of Freud's psychology (superego, ego, id), and selfish genes and memes.

As the examples suggest, I think it is common for paranoid explanatory projects to posit small agents and paternalist projects to posit large ones. The tendency is not invariable, of course, as Satan and the angels attest. And while sometimes there *are* large and kind agents or small and vicious ones at work, the list of examples is intended to suggest that the psychological appeal of such hypotheses often far outruns their empirical warrant.

The transition between styles of explanation in biology was accompanied, I said above, by the exchange of one set of beneficiaries for another. Harmonious groups were replaced by selfish genes. But the new set of beneficiaries acquired too powerful a role, and one tradition of foundational description of evolution devolved into Darwinian paranoia.

My talk of "paranoia" in this context draws on the work of Richard Francis (2004). Francis argues that parts of contemporary biology have come to prize, above all others, explanations in terms of hidden *rationales* for biological characteristics. The biologist is induced to expect that there is some such rationale for nearly everything, and if we cannot find one that is a kind of scientific failure. Francis uses the phrase "Darwinian paranoia" more broadly and less psychologistically than I do. For Francis, pure adaptive thinking in biology itself tends towards paranoia, even without posits of hidden plotting agents. The concept of adaptation has a special intermediate status here, being

---

[7] Sometimes the direction of wagging is explicit: "The whole purpose of our search for a 'unit of selection' is to discover a suitable actor to play a leading role in our metaphors of purpose" (Dawkins 1982a: 91).

useable in both thick quasi-teleological senses and much more minimal ones (Lewontin 1985, Burian 1992). But there is a style of selectionist thinking in biology that I think does not involve any form of paranoia. This is the kind of investigation where someone asks: suppose a population was like this, and such-and-such a mutation appeared, what would happen to it? Thinking this way does not require the idea that genes are "ultimate beneficiaries" of anything.

The reproduction of genetic material is part of the reproduction of cells and organisms. The scrambling of genetic material is one of the consequences of sex. These facts support two kinds of Darwinian description of genes, one weaker and one stronger. The weak kind is supported by the fact that any collection of stretches of DNA located in the right places within organisms can be described as varying, passing on their differences in reproduction, and influencing their chances of being copied. The stronger kind is seen in the special cases where a gene proliferates through a process whose crucial steps do not involve a contribution to organism-level reproduction. Those processes are largely dependent on the machinery of sex. With the scrambling of genetic material comes the possibility of independent action, and selfish genetic elements are part of the cost.

During early and classical genetics, a "particulate" way of thinking about genes was undoubtedly progressive. The positing of Mendelian "factors," and then genes, which remain intact and pure across generations despite combining with other factors inside different organisms, was a huge advance. But as our knowledge gets finer-grained, talk of genes as units is slowly being replaced by talk of genetic "material"—a stuff, not a discrete unit—and by flexible talk of genetic "elements" when the causal roles of particular pieces of this material are under investigation. Known paradigm cases of evolution by natural selection depend on the high-fidelity copying of genetic material, but rather than being the clearest and most fundamental units of selection, genes themselves in most cases are marginal Darwinian individuals.

CHAPTER 8

CULTURAL EVOLUTION

8.1. Populational and Darwinian Models of Culture

[*Limits of populational explanations; two ways of locating Darwinian populations in culture; cross-cutting distinctions concerning levels.*]

One motivation behind many abstract formulations of Darwinism has been to fashion a tool that will be useable in new domains. Cultural change is a prominent example, and the history of thinking about culture in Darwinian terms is almost as long as the history of Darwinism in biology. This history has been one in which successive writers have often seized on quite different parts of the "Darwinian" framework, but there has been a surge of work in recent years that has coalesced around a picture like this: the general capacity for culture in humans presumably has a genetic basis and is a biological adaptation. Social learning, especially by imitation, is crucially important here, as is the capacity for language. But once these capacities are in place, cultural change acquires its own Darwinian dynamic. In some versions of the story, the result is a contest between cultural replicators—selfish memes—in a new Darwinian arena.[1]

Those are the ideas assessed in this chapter. The main aim is to see how various mechanisms that might be found in human culture relate to the Darwinian framework in principle, rather than to say which mechanisms are the most empirically important ones. The term "Darwinian" is used here in the same sense as the rest of the book; it involves change by natural section, change through differential reproduction or replication, though towards the end of the chapter I will look at other senses in which explanations of culture may draw on Darwinian ideas.

The picture that emerges is as follows. Some "cultural evolution" is ordinary biological evolution, via a special set of inheritance mechanisms. Once we go beyond those cases, we encounter something discussed in detail earlier: the possibility of phenomena that have *some* Darwinian features, or approximate

---

[1] For elements of this package, see Dawkins (1976), Cavalli-Sforza and Feldman (1981), Boyd and Richerson (1985), Hull (1988), Durham (1991), Dennett (1995), Tomasello (1999), Jablonka and Lamb (2004), Mesoudi et al. (2004), Richerson and Boyd (2005), Hodgson and Knudson (2006).

them, without having the characteristics associated with paradigm cases. Thus they tend to *invite* a Darwinian description, though this description will apply in a partial way and has the capacity to mislead. But there are also some possible phenomena that are genuinely Darwinian in the adventurous sense seen in recent discussions—phenomena in which culture does form a new Darwinian domain. These are consequences of particular configurations of psychological and social factors. Darwinism is not likely to unify and transform the social sciences, in the way enthusiasts have claimed. But culture has Darwinian roots, and generates additional Darwinian phenomena when circumstances conspire.

The situation can initially be represented in terms of three nested categories. Figure 8.1 shows the relations between three kinds of models or explanations, in both the cultural domain and others.

The "populational" category again refers to—roughly—Mayr's notion of population thinking, discussed in Chapter 1. But when population thinking is brought to bear on culture as opposed to biology, some features of this approach that were obviously applicable in the biological case become more questionable. So the broadest "populational" category in Figure 8.1 is not supposed to cover the entire field of options.

When we embark on population thinking, we treat a system as an ensemble of individual things, which have some degree of autonomy and a significant number of properties in common. We should also know roughly where one ends and another begins. Some collections of things are too tightly integrated to be usefully seen as populations—the atoms in a hemoglobin molecule, for example. Other systems have parts that are too different from each other, and whose roles depend primarily on those differences—a car's engine. A highly structured network with heterogeneous and non-interchangeable parts is a different thing from a population. A riot, in contrast, is a populational phenomenon, as is the mixing of molecules in a gas. Between these clear cases we have intermediate ones—a parliament, an orchestra. They are made up of things with many shared properties and some autonomy, but there are significant asymmetries between

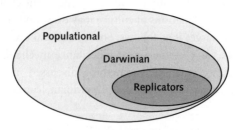

Figure 8.1: Three categories of explanation.

the components as well, and much of what happens is a consequence of these asymmetries in role.

If a hemoglobin molecule is not a population of atoms, what about the collection of cells within an organism? Often in this book I have treated organisms like us as Darwinian populations of cells. But we are mixed or intermediate cases as well. Cells in us have some features that definitely justify a populational approach—they are simple reproducers, in particular. But their organization, and the manner in which they give rise to whole-organism characteristics, are only partially population-like in this sense. The cells in a human are a bit like a parliament, or an orchestra—but one whose members reproduce as part of their orchestral activities.

So although it might initially have seemed obvious that cultural change is a populational phenomenon, in the present sense this claim is far from trivial. Some cultural phenomena are populational in character, and some are not. Cultural phenomena of all kinds *depend* on the activities of individuals who make up populations, but it is possible for a population to generate products that are not best treated in populational terms.

Persisting community-level artifacts like buildings and computer networks, for example, are the consequences of activities of a population, but once they exist their ongoing role is not populational in character. And structures like these, once they become very elaborate, may affect behavior in ways that *reduce* the populational character of social life. Highly structured societies with top-down control are also less amenable to a populational treatment. The ideas that proliferate are those that come from a certain *location* in the society, regardless of their content and local consequences. Reisman (2005) argues that Darwinian models of culture become less applicable as power relations become more asymmetric. I am broadening the claim to one about populational models in general. When a society develops networks of interaction featuring extensive asymmetries in role, it becomes less population-like.

These features, described in different terms, have been the basis for some general rejections of the idea that cultural change is a Darwinian process—Fracchia and Lewontin (1999) is a forceful example. But those anti-population observations should not be made in too general a way. Simpler forms of culture may have a more populational character than more complex forms; an initial populational mode of interaction may give rise to something else.

Suppose next that we are dealing with some phenomenon that is clearly populational. In the domain of culture, these might include changes in patterns of individual everyday behavior (eating, communicating, cooperating). When are such processes also Darwinian? This question is supposed to be answerable using the account given in the previous chapters. We ask whether the entities in question make up a Darwinian population, and are changing as a consequence of variation, heredity, and different rates of reproduction.[2]

In the case of culture there are several ways in which Darwinian populations might be recognized. I will divide these into two main options. The first is the simplest. The entities said to make up the population are ordinary biological individuals, such as people, and culture is treated as an aspect of their phenotype. People have cultural properties (skills, vocabularies, habits), and they vary in these properties. When people reproduce, their offspring often resemble the parents with respect to these features, as a consequence of teaching and imitation. And some people reproduce more than others. The result is evolutionary change.

This is clearly a way in which cultural characteristics can evolve by a Darwinian process. It is not a *new* application of the theory, in fact, but an ordinary one. Darwinism does not require any particular mechanism of inheritance, and here the mechanism is non-genetic. Talk of reproduction and fitness is understood in the usual way, in terms of the production of offspring.

The role of this first option is limited in obvious ways. It cannot capture cases where people copy behaviors from people other than their parents. (It only handles "vertical" as opposed to "horizontal" and "oblique" transmission.) So it may seem that we need a notion of a "cultural parent" as opposed to a biological one. Those you copy, with respect to a particular trait, are your cultural parents with respect to that trait. Individual people are still seen as the members of the Darwinian population, but they are now linked by a non-biological parenting relation.

When we look at phenomena such as accents, expressions, and fashions, this can seem like a promising approach. But it is not really a way of applying *Darwinian* ideas to culture. Reproduction is the creation of a new thing, a new member of a population. When you adopt a phrase from someone you hear talking, you have been changed, but not re-created or reborn. One might reply at this point: so much the worse for a theory of culture tied to the idea of reproduction. That is in many ways the right reaction. But there is a way of handling some phenomena of this kind in a Darwinian way, and that involves moving to the second main family of options.

The second approach is to see instances of cultural variants as making up their *own* Darwinian population, connected by reproduction. Your father's, or your best friend's, Catholicism might be the parent of your Catholicism; his instance is the parent of your instance. The entities in question might be behaviors,

---

[2] We might also have a category labeled "evolutionary" between the populational and Darwinian ones—a category using other mechanisms from evolutionary theory, beside Darwinian ones. I will largely ignore that possibility for the sake of simplicity, though see Richerson and Boyd (2004) on this issue. The topic also arises in the final section of the chapter.

psychological states, or material artifacts. I will use phrases like "instances of cultural variants" in a broad way to include all of these.

This second option does not, in principle, require replicators. Here, as before, I understand replicators as members of a Darwinian population that reproduce asexually with high fidelity, "preserving structure" over many generations of copying. Though this second option need not be presented in terms of replicators, it is common to do so. This is how Dawkins and others develop the hypothesis of *memes*, gene-like entities with replicative power, whose dynamics generate cultural change.

The broad distinction above has initially been developed in an individualist way. Either human organisms make up a Darwinian population passing on cultural phenotypes, or the instances of cultural variants that individual people exhibit make up the population. But the same distinction can be applied at the group level, and both options are possible there as well. It could be argued that human groups have cultural phenotypes that are transmitted to offspring groups (Henrich and Boyd 1998, Sterelny, forthcoming), or that group-level cultural variants themselves (such as forms of political organization) may make up a pool of reproducing entities. So we have two cross-cutting distinctions, one concerning the type of thing said to make up the population, and hence the associated notion of reproduction, and the other concerning the level at which the population exists.

8.2. Reproduction and Causation

[*Group reproduction and persistence; return to the menagerie; formal reproduction by cultural variants; causal directionality.*]

In this section I will look more closely at reproduction, and hence heredity and fitness, within each of the approaches above. The first option treats culture as a set of characteristics of individuals transmitted across generations by non-genetic means. When the population in question is a population of humans or other animals, no special handling of reproduction is needed. This, again, is just ordinary Darwinism.

When this first approach is applied at the level of social groups or communities, extra issues arise. These are familiar from the discussion of collectives in earlier chapters: when do we have genuine group-level reproduction? Darwinian language is often applied to social groups and communities in such a way that the focus is on *persistence* of a group as contrasted with extinction, or growth as opposed to shrinkage. The famous case of the competitive interactions between the Nuer and the Dinka tribes in Africa, often used to illustrate "cultural group selection," is an example (Sober and Wilson 1998: ch. 5). The Nuer and

Dinka are closely related tribes in the Sudan area whose violent competition was closely studied in the early twentieth century. The Nuer prevailed, largely through better large-scale organization of their fighting forces. In this book I treat Darwinian processes involving growth and persistence without reproduction as marginal cases. This is not a stipulative matter, as it shows up in the kinds of explanations that can be given. Origin explanations are particularly important forms of Darwinian explanation, for example, and selection does not figure in origin explanations when selection is a matter of differential persistence only. A group that persists does have the opportunity to transform into another kind of group—in that sense persistence is linked to novelty. But it is a weaker link than the one seen when there is differential reproduction. So "cultural group selection" of a significant kind requires differential reproduction, not just differential persistence, even though the border between these is vague.

The second family of options views instances of cultural variants themselves as making up Darwinian populations. Once put in the language of this book, such a view initially sounds strange. It *is* a strange idea, though this fact can be obscured by patterns of description that reify cultural traits and make things like "ideas" sound more concrete than they are. Though a strange idea, it may sometimes be a useful one, and about half the strangeness has been confronted before, in the form of biological examples.

At the end of the menagerie in Chapter 4 I discussed cases of "formal reproduction." In most biological reproduction, parents contribute in both material and formal ways to offspring. Indeed, the "material versus formal" distinction looks forced and suspicious here almost all of the time (Oyama 1985). But there are special cases where there is a parent–offspring relation of the kind relevant to Darwinism, without the parents contributing materially to the offspring. The cases discussed were retroviruses, prions, and LINE transposons. These are special kinds of scaffolded reproducers, things whose reproduction is highly dependent on machinery outside of them. Because of that dependence, their reproductive abilities are fragile. In slightly different contexts they could not reproduce at all. But they can all, to various degrees, enter into Darwinian processes. Retroviruses, in their nefarious way, are paradigm cases—things whose existence would be quite baffling in the absence of Darwinism. So formal reproduction can be a basis for Darwinian evolution.

These give us a biological model for the idea that reproduction, of the sort relevant to Darwinism, can be a relation between one instance of a cultural variant and another. Suppose you are the very first person to use a turntable as a musical instrument. A few people see or hear you, and do the same thing. The behavior spreads. It is not quite accurate to say that your individual behavior was the parent of theirs, but something close to this is true. They acquired their disposition to do such things as a consequence of the existence of yours.

The simplest cases are those where each new adopter picks up the habit from observing just one person doing it. Then there will be an asexual reproductive lineage of instances of the habit. But this single-parent feature is not essential—to think so is, again, to accept too much of the replicator framework. Heredity of the kind needed for a Darwinian process can exist with more than one parent, as long as the reproduction relation is clear. An example is given in the next section.

These "formal" kinds of reproduction have special features, however. In ordinary cases of biological reproduction, the parent has an overall causal responsibility for the offspring; the offspring does not initially exist at all, and is materially brought into being by its parents. In many cases of formal reproduction there is a pre-existing object whose state or organization is changed, and that object has its own causal properties. The result is a kind of mixed causal responsibility for the fact that something gets reproduced.

The biological cases again provide a model. A prion protein was not mere unformed matter when it encountered the prion that "switched" it. It was a protein molecule with a specific shape, an amino acid sequence, and hence a set of dispositions to respond to other molecules. It happens that the response of some proteins on meeting a prion is to assume the shape of the prion. (Not all prions have parent prions. Some occur spontaneously.) This may be described as the prion *inducing* the protein to re-fold into the prion configuration, but it definitely takes two to tango.

In the biological domain this is a very unusual case. In the case of cultural entities this situation is common. The reproduction of habits, accents, and ideas is largely a consequence of the dispositions of the agent *adopting* the variant. As far as the bare possibility of a Darwinian process is concerned, this does not make much difference. A similar pattern of spread of a trait could be brought about by very different mixes of causal responsibility on the part of donor and recipient. Suppose there is a population of individuals of different colors who each endeavor to paint other members of the population the same color as themselves. If painting-proficiency is affected by color in some way, then a particular color—perhaps red—may spread. We could track a growing tree of instances of red coloring as it moves through the population. The same pattern could also come about through a very different causal process. Now individuals do not impose their color on others, but paint only themselves. They choose their color by observing other members of the population. If one color—red—was more likely to be chosen, it may spread through the population, and the shape of its spread may be the same as in the first case. The differences between the two cases only show up when the situations are perturbed in some way—when we imagine shifts in the policies or dispositions of the agents. In the second case, the "recipient" of a color is causally responsible for almost all of what happens—the inclination to paint, the choice of model, and the function relating the model's color to one's own color. If the recipient changes with respect to any of those policies, the match between the two individuals—and the link in the chain by which red spreads—does not occur.

Most cultural variants spread by something more like the second of these scenarios. The causal directionalities in most biological reproduction are like the first.

Here it is useful to return to some ideas of Griesemer's that were discussed in earlier chapters. Griesemer holds that biological reproduction involves "material overlap." I argued against that view. But Griesemer does discuss processes he calls "copying," which do not involve material overlap. He emphasizes the empirical differences between these cases and the ones in which there is a material contribution. If an entity is to reproduce without emitting a material propagule of some kind, it is highly dependent on surrounding conditions to do most of the work. Biological systems cannot usually trust their surrounds to do this work for them; if they are to reproduce, they need to manipulate matter themselves, and that usually requires taking matter *into* themselves, and re-shaping it with the aid of energy. This is true, but it is also true that sometimes a biological entity *does* find itself able to rely on external machinery to do these things on its behalf, and we see that in retroviruses.

Something of the same situation can be seen with other "selfish genetic elements," even when their reproduction does include a material contribution. Once certain machinery is reliably in place—sex, meiosis, crossing-over—there is an arena within which a special set of Darwinian processes can occur. Similarly, once a population of human agents has the right set of dispositions, instances of cultural variants can proliferate, forming lineages that can be described in terms of reproduction. In the case of genetic elements, this machinery only changes slowly, through evolutionary processes. In the case of cultural objects, the required external "machinery" is much more fragile, and often not present at all. It only exists when the agents apply simple habits of imitation, picking behavioral models and copying them without transformation and customization of the behavior acquired (Sperber 1996, 2000). Much of the time human agents do not behave like that, and as agents integrate more information and modulate their choices, the parent–offspring relation first attenuates and then disappears.

Above I discussed "cultural variants" in a very general way, but different aspects of culture may have better and worse fits to these requirements. Most writers in this area have supposed that the "culture" that Darwinism may help us explain consists in mental structures of some kind, or behavioral dispositions (Richerson and Boyd 2004), and that option has also been the focus of most of the criticism. Sterelny (2006) argues that the best candidates for cultural replicators are not ideas but persisting artifacts—tools, for example—especially in early stages in human history. Artifacts like tools may, in principle, be scaffolded reproducers;

recall the discussion of nests in Chapter 5. But as above, the recurrence and success-guided proliferation of some type of tool is not enough for a Darwinian process. What is required is that each instance of the artifact have a small number of "parent" instances, so that asexual lineages or family trees are formed, with a pattern of heredity across the links.

Introducing a discussion of Polynesian canoes, Rogers and Ehrlich (2008) translate and quote a 1908 passage from the French philosopher "Alain" (Émile-Auguste Chartier), who claimed that as "every boat is copied from another boat," a Darwinian approach to explanation can be taken. Badly designed boats will sink, and hence not be copied, so "it is the sea herself who fashions the boats, choosing those which function and destroying the others" (p. 3417). If this claim about copying is true of boats (or rather, if it *was* true of boats), then we have just what is needed. What is required is not just that one existing boat occupy a special causal role in the construction of the new one. That is compatible with the new boat-builder using the old boat primarily as a spur to tinkering and experimentation. Exact or faithful copying is not needed, however—boats need not be replicators. The causal role seen in an evolutionarily relevant parent–offspring relation lies between these; the parent is causally responsible for the offspring in a way that produces similarity between them. The required relationship could also hold between smaller entities than whole boats. It would be possible for an artifact like a boat to have dozens of parent-artifacts when considered as a whole, but where each affected just one aspect of the new artifact, in a discrete way. (The mast is copied from this boat; the rudder-blade from that one.) Cultural reproduction may be more piecemeal than biological reproduction, and that would not prevent lineages being identified in artifacts at a finer grain. But once general intelligence intervenes in such a way that a vague and disparate set of models all make blended and customized contributions to the new boat, net, or hut, the Darwinian pattern is lost.

8.3. Imitation Rules and Others

[*Dynamics of imitation; relation to Darwinian processes; imitation rules as part of a larger family; selfish memes.*]

Next I will discuss a body of recent work that can be used to illustrate and extend the themes of the previous section. These are abstract models of change in populations due to social learning (Skyrms 2003, Nowak 2006). The models are mostly studied using computer simulations. An idealized population—often situated in a spatial network of some kind—contains individuals who interact with each other, receive payoffs according to the behaviors they produce and encounter, and then update their behavioral dispositions as a consequence of their experience. The models use a number of different rules for this behavioral updating. One rule that is particularly interesting in this context is "imitate your best neighbor" (IBN). Each individual is assumed to interact at each time-step with some small set of immediate neighbors (perhaps the north, south, east, west neighbors on a lattice, as in Figure 6.2). Each individual can also observe the overall payoff received by each of these neighbors, as a consequence of the neighbor's own total set of interactions. At the next time-step, each individual produces the behavior that was produced by the neighbor receiving the highest payoff on the previous time-step. (Or it produces the same behavior again, if the individual did better than any of its neighbors.) So each individual in the population is constantly shifting its behaviors in accordance with what has been locally successful.

In Skyrms' models, a combination of local interaction and updating by an IBN rule was shown to lead to strikingly cooperative or "pro-social" profiles in these idealized populations. One example is the "Stag Hunt" game. Here each individual must choose at each time-step between "hunting hare," a solitary activity, and "hunting stag" in a cooperative way. The payoffs are affected by what one's partner chooses—interactions are structured in pairs. If you choose to hunt cooperatively, you receive a payoff of 3 units if your partner also cooperates, but zero if he does not. If you choose the solitary option, you receive 2 units no matter what your partner does. At each time-step, each individual plays the game separately with each of its neighbors, and its overall payoff is a sum of each of the particular payoffs that result.

Skyrms found that when this game is played on a lattice with an IBN rule, the population almost always evolves to a cooperative outcome. Similar results were found with some other well-studied games, such as "divide the dollar." Here two agents issue demands for how a fixed windfall is to be divided. If the proportions claimed by each agent sum to 100 percent or less, each agent receives their demand. If the demands sum to more than 100 percent, neither receives anything. In a simplified version of this game (with fewer options than usual), Skyrms found that with local interaction and IBN, populations almost always evolved to a situation where all individuals made the "fair" demand of 50 percent.

Skyrms offers these results as a sketch of how aspects of our "moral sense" might have evolved, and how cooperative behaviors can remain stable. The explanation works by showing that cooperative behaviors function, to a surprising extent, as "attractors" in certain kinds of population dynamics.

These models are clearly populational ones, in the sense discussed above. That itself can be the basis for criticism; moral learning, and our resulting moral intuitions, may be examples of elements of culture that do not work in a populational way, because of the importance of power structures and other asymmetries in social roles (Levy, forthcoming). But let us bracket that problem, and assume they do have the potential to explain something significant. How do such models relate to the idea that cultural change is a Darwinian process?

This is not a scenario driven by births and deaths, in any obvious sense. Instead it is more naturally interpreted as one in which agents persist and change their state in response to experience. (Births and deaths *could* occur but are irrelevant.) If an IBN rule is being followed, however, the model can be interpreted as Darwinian in the sense of the second option described earlier, in which instances of cultural variants themselves form a Darwinian population. When an individual copies its next behavior from a particular neighbor who did well on the previous time-step, one individual's behavior at time t is the formal parent of another individual's behavior at $t + 1$. A particular instance of a behavior might, through successive events of imitation, be the ancestor of a branching tree of descendant behaviors, spreading through the population. Each behavioral instance is transitory, but if successful it may be causally responsible for other behaviors of the same kind. Behaviors themselves in this system are replicators.[3]

So *given* the assumption of a particular rule that all the agents are following, behaviors can form parent–offspring relations and be part of a Darwinian process. If agents start to respond to their experience differently, the parent–offspring relation collapses. These alternatives need not be anything particularly creative. The IBN rule is one of a family of plausible rules available to agents of this kind. A simpler imitation rule than IBN is "copy the common." Here an agent assesses the local (or perhaps global) frequency of various behaviors and produces, on the next time-step, the most common. This is a less "smart" imitation rule than IBN, and it also lacks the Darwinian features of IBN, both because it is not success-driven and also because any given behavior will not have a single "parent" behavior on the previous time-step. But copy-the-common rules are frequently employed in actual populations (Richerson and Boyd 2004).

Skyrms himself compares the IBN rule with "best response" rules of the kind often used by other game-theorists. A simple example of such a rule would have an agent produce on the next time-step the behavior that would have been the most appropriate overall response to the behaviors produced by the individual's neighbors on the previous time-step. A best-response rule is "smarter" than IBN, as it requires not just tracking payoffs actually received by others, but tracking which behaviors *would* do well in various circumstances. Interestingly, Skyrms found that a best-response rule produced less cooperative outcomes than IBN in

---

[3] It would be possible to do a "coalescent" analysis of the pool of behaviors present at a time; you could determine the average time to a common ancestor for two randomly chosen behaviors, for example.

his model. In a Stag Hunt, that means that individuals are being too smart for their own good, as they do best if all cooperate. A best-response rule also lacks the Darwinian features of IBN. It is success-driven, but it does not lead to a situation where particular behaviors form parent–offspring lineages.

Once we are focusing on this lineage-forming property, we see that IBN is quite a special rule. It is not the only one that can produce Darwinian results, though. Suppose an individual determined their next behavior not by imitating a single neighbor, but by blending or averaging the behaviors of its two highest-scoring neighbors. (If the behavioral choice is a binary one, such as cooperate versus defect, then this might involve adjusting the probabilities of each behavior.) Now each behavior instance has two parents; this is sexual reproduction in the pool of behaviors. As discussed in Chapter 2, this can be a Darwinian system even though there is no replication.[4] Behaviors will still show heritability. Behavior instances will not form branching asexual lineages, as in the simplest form of IBN, but will form networks like human "family trees."

What we have found here is a family or space of possible update rules in which an individual's behavior is some function of what it is exposed to. Having one's behavior be a *function* of the attributes of one's neighbor(s) is not the same thing as having one's behavior be a *copy* of some neighbor. The latter is a special case of the former. Some of these update rules have a Darwinian character, and some may even allow us to identify replicators. But there are many other ways, beside Darwinian ways, in which a population's past can feed forward to affect its state in the future, in a way that involves aggregations of local individual responses.

To follow this argument up, I will describe the situation more formally, by representing some of the family of update rules with a single formula containing adjustable parameters that make the resulting process more or less Darwinian. Suppose an individual uses its experience to set the value of some behavioral characteristic Z for the next time step, $Z(t + 1)$. Z is a continuous variable (though it might be the probability of making a binary choice). Assume the individual has n neighbors, where neighbor i's behavior at time t is represented as $X_i(t)$. $Z(t + 1)$ may then be a function of the behaviors of all the neighbors at t, their payoffs (W_i) at t, along with the individual's previous state $Z(t)$ and payoff $w(t)$. So in general, $Z(t + 1)$ is some function of the following variables: $(X_1(t), X_2(t), \ldots, X_n(t), W_1(t), W_2(t), \ldots, W_n(t), Z(t), w(t))$. This makes possible a vast range of rules, some of which can be represented like this:

$$Z(t + 1) = uZ(t) + vX^*(t) + (1 - u - v) \sum_{i=1}^{n} X_i(t)/n \qquad (1)$$

---

[4] As described here, this inheritance system will lead to the loss of variation. For discussion and modeling of the cultural inheritance of behaviors without replicators, see also Henrich and Boyd (2002).

Here $X^*(t)$ is the behavior of the neighbor with the highest payoff at t, or the behavior of the focal individual at t if its payoff was higher than any neighbor's. The "weights" u and v are positive and sum to some number between zero and one (inclusive).

The idea is that any individual's new behavioral choice can be sensitive to (i) what it did last time, (ii) the recent success of behaviors exhibited by neighbors, and (iii) the local prevalence of those behaviors. An individual can give some role to inertia, some to tracking what has recently worked, and some to doing what is common. The u and v parameters reflect how much weight is given to each factor. So when $v = 1$ we have the IBN rule, when $u = 1$ the individual never changes, and when both are zero we have a version of "copy the common." But the weights can also take intermediate values.

In this family of rules, the only way an individual can directly respond to success is by attending to $X^*(t)$. The other influences are inertia and conformity. But this process of behavioral change by individuals could operate "on top" of a Darwinian process involving biological reproduction. An individual will then be born with an initial behavior, and will update it according to some specific rule—some particular settings of u and v. These weights could then evolve across generations. When does an individual do best to stick with a behavior inherited from its parents ($u = 1$), and when does it do best to adjust in the light of its own experience? If it should adjust, how quickly should it do so? In a very noisy world, individuals may do well to keep u close to 1, and hence update only slowly in the light of what is common or what is successful.[5] It would be interesting to model this system in detail.

In this scenario, Darwinian processes give rise to a rule for social learning, and that rule may or may not have a Darwinian character itself. If v evolves towards one, the dynamic in the pool of behaviors will become Darwinian; behaviors themselves will come to form parent–offspring lineages. If v evolves away from one, the pool of behaviors will not have these Darwinian qualities. Darwinian biological processes may produce Darwinian or non-Darwinian forms of social learning.

The range of possibilities in formula (1) is also, as we have seen, the tip of the update-rule iceberg. The formula does not include a best-response option, blends of multiple successful neighbors (perhaps weighted by how successful they were), and many others. There will usually also be a place for simple trial-and-error learning (which has its own partially Darwinian character).[6] So evolution can build agents who use social experience to influence their choices in a number of

---

[5] This gives the model a link to the larger literature on the evolution of learning: Stephens (1991), Bergmann and Feldman (1995), Godfrey-Smith (1996), Kerr (2007).

[6] See Campbell (1974), Dennett (1974, 1995), and Hull et al. (2001).

ways. It is a striking fact that some of these ways, including IBN, can generate a new Darwinian population in the pool of behaviors themselves. But evolution may or may not build such agents. And it may build them initially and then build something beyond them—suppose biological evolution produced a sequence of successively "smarter" rules in a population: first copy-the-common, then IBN, then a best-response rule. The pool of behaviors is initially non-Darwinian, becomes Darwinian, and then becomes non-Darwinian again.

I will finish this section with some further discussion of the idea of memes. The view described above allows the possibility of some meme-like cultural phenomena. By this I mean cultural variants that form lineages of copies—discrete cultural entities that are reproduced in a high-fidelity and asexual way. Once this is seen as one possibility among many, the questions become how common it is and why it should arise. It may be that particular cultural circumstances produce phenomena of this kind, or at least something approximating them. Above I emphasized the simplicity of the individual-level psychological dispositions that generate such things. Once people combine too many sources of information and manipulate that information too intelligently, the phenomenon will vanish. But there may also be cases where despite much available smartness, gains can be made by simplifying the transmission of cultural variants. Perhaps these include situations of overload—as the soup of cultural possibilities becomes so complex as to be overwhelming, there is an inclination to adopt cultural variants in a discrete and simple way. A hint of this is seen in the increasing power in contemporary Western culture of discrete cultural fragments—brands, icons, compactly named political identities, and social profiles. This might, I realize, reflect no more than our need to compactly *describe* some things which may themselves be heterogeneous and noisy. And it is a long way from the extreme of simplicity of transmission that I associated with cultural replicators above. But might it be that people become willing to adopt a *more* discrete-and-combinatorial approach to cultural choices, as well as cultural description, as the range of options becomes unmanageable?

So far I have not mentioned the feature that made the meme hypothesis so startling when Dawkins introduced it. This was not the idea of an evolutionary theory of culture, or even a gene-like "particulate" one, but the idea of cultural particles which are *selfish*—cultural entities harboring evolutionary interests, and whose interests are not entirely aligned with ours. If someone wonders why a peculiar religious idea survives, for example, this is to be explained in terms of the idea having properties that can be seen as effective strategies for replication in an environment of human agents like us.

The criticisms made of agential views of evolution in earlier chapters apply here as well. We are being induced to see a complex phenomenon through an agential lens. This, I argued, switches on a particular part of our psychology. And as the agents we are invited to believe in are small and potentially unhelpful, the result is another outbreak of Darwinian paranoia. The phenomenology of explanation—the "Aha!" feeling—also plays a role here. We *feel* that we have achieved a special kind of insight when we can assign an event to a hidden agenda. The "Aha!" of an agenda marks a sense that understanding is complete.

There is also another aspect to this situation, which mirrors something encountered in the case of the gene's eye view of biological evolution. In much talk of memes, and particularly in Dennett's development of the idea (1991, 1995), a negative point is being made. What is being opposed is a traditional "rational choice" model of how cultural variants come to be adopted by human agents. In some parts of my discussion above, I made it sound like a *free choice* whether an agent imitates in a simple way or does something more complicated; the agent surveys the options and makes a choice based on which will serve his or her goals. For some, including Dennett, this is a highly inaccurate view of human decision. We cannot step back as rational agents and survey the cultural fragments presented to us; it is more accurate to say we *are*, at any time, just a collection of these fragments, and the collection is organized in such a way that some new fragments make their way in to become part of us and others do not. So the selfish-meme view does not just posit copying of cultural variants, but also serves to oppose views that treat the human agent as "a sort of punctate, Cartesian locus of well-being" (Dennett 2001: 70).

In response: perhaps traditional rational-choice pictures of the human agent do fail. Then they should be replaced, and we should give a new theory of the ways that ideas and habits are "taken on board." But there is no need to come up with a new set of hidden *agents*, in developing this story. The breakdown of the unitary agent might leave an initial vacuum, but that does not mean some *other* agent-like entity has to fill the vacuum. Or at least, we should not posit new agents without evidence that directly supports such a view. There is no such support to be found in the evolutionary theory of culture. The most that such a theory could do is describe the existence and dynamics of a new population of reproducing things, a pool of cultural entities that form lineages and evolve.

Agential description of genes, as discussed earlier, was part of a shift away from talk about one kind of "beneficiary" of evolutionary processes to talk of another. Harmonious groups were replaced first by selfish individuals and then by selfish genes. Once we are *inside* an agential mode of description, it becomes natural to mark the arrival of a new causal model with a new beneficiary. And such talk does shade into talk of effects on fitness that can be literally interpreted. But agential talk about evolution is always no more than a metaphorical gloss on the real populational phenomena, and a potentially misleading one, whichever beneficiaries are chosen.

There is also an irony to note here. Defenders of replicator-based views of evolution and culture often offer deflationary psychological treatments of religion (Dawkins 2006, Dennett 2006). They attribute religious interpretations of experience in part to our natural tendency to attribute complex events to hidden agents. I agree that this is probably part of the explanation for the prevalence of religious ideas in human history. But the same thing is going on, in a flip-side small-agent form, within selfish-replicator treatments of evolutionary processes.

8.4. Conclusion

In this chapter I have recognized several mechanisms, phenomena, and possibilities that amount to Darwinian processes in the cultural domain. These have been organized with two cross-cutting distinctions. The first distinction concerns two kinds of entities that might make up a Darwinian population. One option is that the population is made up of ordinary biological reproducers of some kind, and cultural traits are characteristics of these individuals, aspects of their phenotypes. The second option is that instances of cultural variants—behaviors, ideas, artifacts, words—can make up Darwinian populations themselves, engaging in formal reproduction of some kind.

The second distinction concerns levels. Both the possibilities above may occur at (roughly speaking) the level of individuals and groups. With respect to the first option, it is obvious what that means. Individuals or groups can be the biological reproducers in question. With respect to the second option, the distinction between levels has to do with the nature of the cultural variants that make up the population. The variants may be individual-level habits (such as saying "cheers"), or group-level habits (such as voting by secret ballot). The distinctions between levels are less straightforward within the second option, as there are many cultural characteristics that might be best seen either as individual-level dispositions or as features of a community. Linguistic attributes are a famously vexed example.

With respect to all these options I have argued that quite specific conditions are required for a cultural process to be Darwinian. It might be thought at this point that things are being narrowed in an implausible way. For example, I said that only simple kinds of imitation will allow instances of cultural variants to form Darwinian populations. As soon as people handle their choices in a more complex way, blending models and influences, reproductive lineages disappear. In response, it might be argued that the world is full of phenomena that look Darwinian but will not fit this narrow set of requirements; surely the way the world fills up with laptops looks very much like the way it fills up with rabbits. But mere *recurrence* of cultural variants is not enough, or even recurrence with some causal role for previous instances—recurrence conditioned by earlier tokens of a type. What is needed is a particular kind of causal responsibility on the part of past instances, a kind that leads to reproductive lineages and the possibility of heredity. This need not be found at the most coarse and obvious grain of analysis—small features of artifacts might form lineages even when the wholes do not—but the requirement is a strong one.

The sense that there somehow *ought* to be a larger role for Darwinian concepts here may also arise from the fact that many processes count as populational, in the present sense, but not as Darwinian. An example is the maintenance and spread of a behavior through a conformist bias ("copy-the-common"), or a best-response rule. I argued in the previous section that these might initially *look* just as Darwinian as a rule like "imitate-your-best-neighbor," but in fact they are not. One rule leads to parent–offspring lineages, and the others don't. When assessing the utility of the broader category of "populational" mechanisms in culture, there is no need to take the Darwinian cases as primary or force others into a Darwinian mold. Sometimes it is thought that Darwinian processes *must* be primary, as only they can explain "adaptation" in a population. But this argument has little bite in a cultural context, where we are dealing with intelligent agents who can accumulate skills and information by a variety of means.

There are also broader senses in which cultural processes might be said to be "Darwinian." Of all the work done on complex human cultural traits under a Darwinian banner so far, perhaps the most empirically informative has been work showing that modern biological methods of "tree reconstruction" can be successfully applied to linguistic change (Gray and Atkinson 2003, Gray et al., forthcoming). This work has shown a capacity to answer questions both about languages themselves and about movements of the populations that speak them. This is the application of a more zoomed-out, "macro-evolutionary" side of Darwinism. The applicability of phylogenetic methods to language change does not give us reason to believe that, despite appearances, each utterance of a word is reproduced from a single or small number of "parent" utterances. Rather, it shows that the tree-like structures generated by large aggregations of low-level Darwinian processes can also be generated by aggregations of different kinds of lower-level events.

Even when understood in the narrow way defended here, Darwinian processes within the domain of culture may have had significant roles in human history. Such processes might have occurred only rarely, but in pivotal periods and contexts. Tomasello (1999) has argued for a crucial role for individual-level imitation learning in the early steps that humans took down their strange and unprecedented evolutionary path. Others have argued that such processes would become especially powerful when groups are the units transmitting and refining cultural characteristics, leading to the beginnings of elaborate practices of human cooperation (Bowles and Gintis 2003). And as human agents evolve their psychologies and social lives, they may from time to time hit on the combination of features that generate Darwinian processes at the level of cultural variants themselves. But the Darwinian possibilities in this area occupy one region in a larger space. Rather than cultural change being a special case of a general Darwinian process, Darwinian evolution—with reproduction and heredity—is one way among several in which a population can have its past feed into its future.

This analysis of cultural evolution exemplifies the general themes of the book. My aim throughout has been to apply some central Darwinian concepts in as direct a way as possible, and to extend them in as straight a line as possible. In some areas this leads to a simplification of issues—this is seen with questions about levels of selection. Elsewhere it generates new pressures, especially around the concept of reproduction. And this "straight line" pursuit of one set of Darwinian ideas leads us, I've argued, out along the tangled branchings represented in Darwin's other great idea, the genealogical interpretation of the tree of life.

APPENDIX: MODELS

The heading of each section indicates (in brackets) the section in earlier chapters where the issues discussed arise. The exception is the final section, which is free-standing.

A.1. Equations for Change (2.1)

The idea of a Darwinian population is treated in this book as describing a "set-up," a way in which things can be configured. But the importance of these configurations comes from the fact that they behave in distinctive ways. The knowledge we have of these behaviors largely takes the form of a patchwork of models. This section surveys some ways of representing change by natural selection in equations, emphasizing the pictures of evolution underlying the formalisms.

I will compare three kinds of representation. The first is a family of models which describe evolution as change in the frequencies of types. These can often be applied over multiple time-steps; their output can be treated as input for another round of change without the need to add further information (a "dynamically sufficient" model or a "recursion"). These include many genetic models, the "replicator dynamics," and some models in evolutionary game theory.

The simplest model of this kind is a model of an asexually reproducing population with discrete generations, assuming no mutation, migration, or drift. Assume there are types **A** and **B**, with frequencies p and $(1 - p)$ respectively. The symbol "W" will be used for fitness-related properties of various kinds—it will be defined slightly differently several times. In the first equation W_A and W_B represent the average number of offspring produced by individuals of the **A** type and **B** type respectively. Then p', the new frequency of the **A** type after one generation of change, can be calculated as:

$$p' = \frac{pW_A}{pW_A + (1 - p)W_B} \tag{A1}$$

The denominator of (A1), mean fitness, can be symbolized $\overline{W}$, and then $p' = pW_A/\overline{W}$. If the fitnesses are either constant or functions just of p, the output can be used as input for a new round of change, so the analysis can be extended over many time-steps. This simplest case can be extended in various directions. One variant is to give a model using continuous time, in which births and deaths occur constantly. The fitness parameters now represent the *per capita* rates at which individuals of a given type contribute to the population by reproducing and dying. Then change in frequency of **A** can be represented as $\frac{dp}{dt} = p(W_A - \overline{W})$. This is the "replicator dynamics" (Taylor and Jonker 1978, Nowak 2006). That term is also sometimes used for the model in (A1), in which case discrete and continuous replicator dynamics can be distinguished.

A second way the model in (A1) can be extended is to introduce diploid individuals and sex. Now we track both the frequencies of alleles (*A* and *a*) at one locus, and the combinations formed by sexual reproduction—genotypes *AA*, *Aa*, and *aa*, with fitnesses W_{AA}, W_{Aa}, and W_{aa}. The frequencies of the *A* and *a* alleles are *p* and *q* respectively. The fitnesses can be interpreted as a combined measure of the chance an individual of that type has of surviving, and of the number of gametes it then produces that go into the next generation (Roughgarden 1979: 28). Assuming random mating (union of gametes), discrete generations, a large population, and no mutation or migration, the formula for change becomes:

$$p' = \frac{p(pW_{AA} + qW_{Aa})}{(p^2 W_{AA} + 2pqW_{Aa} + q^2 W_{aa})} \tag{A2}$$

The denominator is again a mean fitness ($\overline{W}$). The model can also be extended to two genetic loci and beyond.

A quite different representation of change is the "Breeder's equation": $r = h^2 s$. Here *r* is the "response" to selection, defined as the difference in the mean of some quantitative character after selection, and the mean before selection; h^2 is heritability and *s* is the strength of selection. In the simplest case (used in a derivation by Roughgarden 1979: ch. 9), this "strength" is the difference between the mean of the individuals in the parental generation who breed, and the overall mean in that generation.

The breeder's equation is based on an underlying genetic model, assuming many genes with small effects. It is designed to be used where the genetic basis for a trait is complex and unknown. But it can also be understood even more abstractly, as heritability itself can be understood in a way that does not assume the presence of genes (Section A.2).

The breeder's equation itself also does not require that the population can be categorized in terms of types, only that individuals have values of a quantitative character. The equation is valid only over a single time-step; even assuming that *s* is constant over generations, the heritability will usually change as the population evolves, and this change is not tracked in the equation itself. Even within this constraint, the equation in most cases applies approximately rather than exactly (Heywood 2005).

The equation embodies a very intuitive picture of evolution, however, one that affects many verbal discussions: fitness differences are not sufficient to generate change if the characteristics of the fit are not "transmitted" to some extent. Heritability is a sort of "channel" between the generations, which may be clear, noisy, or altogether lost (when $h^2 = 0$).

A third approach uses the "Price equation" (Price 1970, 1972, 1995, Frank 1995). This has in common with the breeder's equation the fact that it applies only over a single time-step or (more exactly) interval, and it does not require the presence of persisting types in the population. It applies exactly, however, unlike the breeder's equation. My discussion of Price here and in the next section draws on Okasha (2006).

Assume an ancestral and a descendent population whose individuals can each be described with respect to a quantitative characteristic, and assume a relation (interpreted as reproduction) linking individuals across the two times. Change is again represented as a consequence of a combination of fitness differences and heredity (in a general sense). One version of the equation is:

$$\Delta \overline{X} = Cov(W, X) + E(W \Delta X) \tag{A3}$$

Here X is a quantitative character and $\overline{X}$ is its mean at the start of the interval. $\Delta \overline{X}$ is defined as $\overline{X}_o - \overline{X}$, where $\overline{X}_o$ is the mean at the end of the time interval. W is another slightly different measure of fitness: the number of descendants an individual in the parental generation has, divided by the average of those numbers. Each individual in the parental generation is characterized by its X_i and W_i, its phenotype and its fitness; also by X'_i, the average X value of its offspring; and by ΔX_i, its value of $X'_i - X_i$. (The subscripts are omitted from (A3).) Then $Cov(W, X)$ is the covariance in the population between X and fitness. $E(W \Delta X)$ is the average of the products of the W and ΔX values.

In the breeder's equation, heritability (h^2) was used to measure the extent to which fitness differences in one generation have consequences for the next. In the Price equation (A3), heritability does not appear. Rather than heredity being treated as akin to a "channel," the Price equation divides things up differently. The first term describes how change *would* occur if there was *perfect* transmission of character across the time interval, and the second term adds a correction for any "transmission bias."

The Price equation can be used to describe change in frequency of a type (by suitable choice of X), but it can also be applied to a population of individuals treated as unique. Some see this focus on individuals as an important part of the mindset underlying the equation (Grafen 1985), and I think this is true in conceptual as well as technical respects. The Price equation is suited to the view I defended earlier as "evolutionary nominalism": grouping individuals into "types" should be optional in evolutionary description. More precisely, evolutionary theory should allow that its key theoretical ideas should be applicable regardless of the "grain" with which a system is described. One description of a population might group them into a small number of types; another might use a finer "mesh" in its classification scheme, and hence recognize a larger set of classes with fewer individuals inside them. A third might be so fine-grained that no two individuals fall into the same category at all. Initially this might seem to make attempts at description collapse, but that is not so. Unique individuals may be more and less similar to each other, more or less close with respect to some metric.

The underlying model here is also, in a sense, temporal rather than generational. The equation can be applied even if none of the parents reproduce, but some survive over a time interval and some do not (Rice 2004). This will be discussed in more detail in the final section below, where a generalization of the equation is presented.

The Price equation, unlike the genetic models discussed earlier in the section, is not *idealized*, in the sense of containing deliberate simplifications. Its use might involve idealizations on a particular occasion, but an analysis with the equation does not work by imagining things like random mating and constant fitness values. Instead it takes a particular case of change, either assumed or predicted, and represents the change by breaking it into parts. This is related to the fact that it does not work as a recursion, something whose output can always be plugged back in as new input to the same equation.

Here I have discussed simple equations representing short-term change, emphasizing the different idealizations they make and their underlying pictures. Models bring with them ways of categorizing things and default assumptions, which are often best made clear via contrasts (Winther 2006). I have presented the three formalisms as separate, but they can be connected in various ways—Price equations can be re-expressed so that heritability appears, for example (see Section A.2 below), and some authors discuss ways in which the Price equation can, with additional assumptions, function as a recursion (Frank 1998). There are also attempts to give single equations with a more ambitious role—representations of the overall tendencies in evolution, especially with respect to adaptation and the maximization of fitness (Fisher 1930, Grafen 2007).

A.2. Heritability and Heredity (2.2)

All treatments of evolution by natural selection include a requirement for the inheritance of traits. Replicator views require the reliable transmission of structure. I argued that this is not needed. What is relevant instead is a population-wide measure of parent–offspring similarity. Some summaries use a comparative criterion: parent and offspring must be *more* similar than other pairs of individuals (Lewontin 1985: 76; Gould 2002: 609). I will return to these below.

What often appears in formal models is a *covariance*. (The covariance between variables X and Y, for n paired measurements, is

$$\frac{1}{n} \sum_{i=1}^{n} (X_i - \overline{X})(Y_i - \overline{Y})$$

where $\overline{X}$ and $\overline{Y}$ are the two means.) Covariance between parent and offspring is used in heritability measures, in particular. Covariances require a trait with a mean value in each generation. This might be an obviously quantitative trait like height, but might also be the probability of producing a given behavior, or a characteristic scored as one for the presence of a trait and zero for its absence.

I will look more closely at heritability. There is a family of heritability concepts (Jacquard 1983, Downes 2007). Some assume a causal model of inheritance that includes genes or something similar to them. But it is also possible to approach the idea in a more minimal way, aiming only to represent predictability relations between parents and offspring. Heritability can then be measured as the slope of the linear regression of offspring character on parental character (Roughgarden 1979: ch. 9). That slope is the covariance between parent and offspring values for the character divided by the variance of the parent values. When there are two parents, their average (the "midparent" value) may be used.

How good is heritability when used to express the inheritance requirement in summaries of what is needed for evolution by natural selection? A variety of problem cases do arise. They arise because heritability is *so* abstract a concept; it throws so much information away. As a result, we may have fitness differences and heritability, but where the details of the inheritance system and the fitness differences conspire in a way that results in no net change.

A simple example was given in Chapter 2. Another has been introduced by Brandon (unpublished, discussed in Godfrey-Smith 2007a). This is a simple case in which there are fitness differences, high heritability, but no change across generations because of a "bias" in the inheritance system that exactly counters the fitness differences. As Brandon says, although heritability is identified with the slope of a regression line, a regression analysis gives us two parameters, the slope and the intercept with the vertical axis. The "bias" shows up in the intercept. So if heritability is understood as a regression slope, then at least one extra parameter needs to be taken into account when using heritability and fitness to predict change.

A third case can be developed by making use of the fact that heritabilities are usually calculated in a way that takes the entire parental generation into account, regardless of fitness differences. Imagine an asexual population that contains variation in height. There is a positive covariance between height and fitness. There is also a positive covariance between parental height and offspring height. But there is no change across generations. This is because although taller individuals have more offspring on average, and taller individuals have taller offspring on average, the taller individuals with the high fitness are not the *same* tall individuals as those that have taller offspring. The high-fitness tall individuals are not the tall-offspring tall individuals. Here is a simple numerical example. There are just six individuals in an asexual population. Three are short, with a height of one meter, and have one offspring each whose height is also one meter. Two are tall (two meters) and have one offspring each, where the offspring are two meters. One individual is two meters tall and has seven offspring—very high fitness—but his offspring are variable in height. Four of them are two meters and three are one meter tall. Then we get the same population statistics back with a larger overall population size. One response to this "heritability fails in the fit" case is to understand heritability in a fitness-weighted way, an approach defended in general terms by Heywood (2005). Others will be discussed below.

The example used in Section 2.3 was one of stabilizing selection. That case, and the two above, have in common the fact that the inheritance system would produce change *alone* if there were no fitness differences. So we might respond to those cases by saying that the aim of a summary of evolution by natural selection is not to say when selection will produce change *from what we had before*, but to say when selection will make a difference *from what would have happened without it*. That makes sense also when we think about the possible intrusion of factors such as migration.

Another case of stabilizing selection, a sexual case, would not be handled by that reply. This is a case of heterozygote superiority with respect to fitness but not with respect to phenotype. Assume the phenotype in question is height. An intermediate height is favored by selection and produced by a heterozygote (*Aa*) at one locus, resulting in a stable equilibrium of gene frequencies. There is a tendency for short individuals to produce short individuals and tall to produce tall, even when the population is in the equilibrium state. There are fitness differences between individuals in this equilibrium state. Yet there is no change. Suppose the population at the start of a generation has genotype frequencies of 0.25 *AA*: 0.5 *Aa*: 0.25 *aa*. Exactly half of the homozygotes of both kinds do not survive to breed, and all the heterozygotes survive to breed. So the pool of gametes contains a 50/50 mix of *A* and *a* alleles. If mating is random then the new generation will again have genotype frequencies of 0.25 *AA*: 0.5 *Aa*: 0.25 *aa*. This is not a case in which the fitness differences act to counter some change that the inheritance system was tending to produce "on its own." If there had been no selection at all (and random mating) the new generation would have had the same genotype frequencies. Yet this is a case where parent phenotype predicts offspring phenotype to some extent, and parent phenotype predicts fitness as well.

This is a case where phenotype is heritable but *fitness* is not heritable; Lewontin noted these cases when formulating his 1970 summary, and this is not a counterexample to that formulation. But requiring fitness to be heritable, rather than phenotype, brings other problems. This move has the consequence that if there are no fitness differences in some particular generation, then the transition *from* the previous generation *to* that new one cannot count as due to natural selection, even if there was extensive change in producing the new generation, and (for example) the pale ones in the old generation all died because they were eaten by birds and color was highly heritable.

Here is another interesting case, due to David Haig (personal communication) and developed in the context of modeling birth weight. Suppose that individual *i*'s phenotype is determined by the function $X_i = T_i + E_i$, where the variable T represents genotype and can be in three states (1, 2, or 3), and E represents environment, also with three states (1, 2, 3) that are equally probable. There is perfect asexual inheritance of T. Fitness, however, is determined by the function $W_i = E_i$, and hence is independent of T. Then X will be heritable and correlated with fitness, but all the fitness differences among individuals are due to the environment, and are uncorrelated with genotype. As a result, there will be no evolutionary change despite heritability and fitness differences.

The problem comes from the fact that environment here is a common cause of fitness and of phenotype. There are several ways of responding to this case. First, one might simply claim that it is a requirement for evolution by natural selection that fitness causally depend on phenotype, as opposed to merely being associated with it. This is only a partial answer. Second, in Haig's case there is a role for something like the "biased" inheritance discussed above in the Brandon example. If we consider only the fit individuals within each genotypic class, we find their offspring are biased downwards with respect to phenotype (and fitness). Third, it is a case handled by Lewontin's 1970 formulation, as fitness is not heritable even though phenotype is.

All of those problem cases can be understood in a different way by using the Price equation. Okasha (2006: sec. 1.5) argues that by recasting the Price equation so that a heritability term appears explicitly, we can see that traditional three-part recipes using heritability are generally accurate for predicting change *except* in cases where either, or both, of two additional effects are present. One effect is a role for the Y-intercept in the regression line used for calculating heritability, which was mentioned above. The other is a covariance between an individual's fitness and the "error" or deviation found when that individual's offspring's phenotype is predicted using that regression line. Okasha noted this second effect as an abstract possibility without giving an example; Haig's case is a moderately realistic case where this feature is present. The "heritability fails in the fit" case above also has this feature; high-fitness individuals have their phenotype badly predicted by the regression line defining the heritability.

The cases of stabilizing selection receive a different analysis from this point of view. The fitness term in Price requires a *covariance* between character and fitness, which is absent in stabilizing selection. So initially, it seems the equation does not recognize the fitness differences at all. But the stabilizing selection cases could be re-analyzed by treating *deviation from the mean height* as the character X being analyzed, rather than height itself. Then we have a negative covariance between X and fitness, and a transmission bias (second term) that counteracts it.

Some summaries express the inheritance requirement in a comparative way, as I noted above. Lewontin (1985) required that "individuals resemble their relations more than they resemble unrelated individuals and, in particular, offspring resemble their parents." There are several ways of interpreting these criteria, but here is one: the average difference between parent–offspring pairs is smaller than the average difference between individuals of different generations. Without taking fitness differences into account, each parental individual is associated with its phenotype X, and also an X' value, the average phenotypic value for its offspring if it has any (see the discussion of Price above). Deviations are squared. When both tests can be applied, this comparative criterion and a covariance criterion for heritability coincide: the difference between the average squared deviation across individuals of different generations and the average squared deviation across parent–offspring pairs is proportional to the parent–offspring covariance. The comparative criterion can be applied in some cases where the covariance criterion cannot, however. Suppose there are many qualitatively different types in both generations, with reliable transmission of type but some probability of mutation. The probability of a "match" in type is high across parent and offspring, lower for other pairs of individuals. There are no mean values defined within each generation, hence no departures from the mean and no covariance, but the comparative criterion can be applied (scoring each pair with 0 for a match and 1 for a failure to match). To use the covariance test we need to redescribe the population, so that all the individual X-values within each generation are scored on a numerical scale.

In the light of all these cases, how should we think of the heredity requirement in descriptions of evolution by natural selection? A Darwinian process requires that parents produce offspring who are similar to them. Whether a case of parent–offspring similarity is evolutionarily relevant depends on the statistical profile of the whole population. So "similarity" is vague, but the population-level models that describe the situation are not. Slight similarities are often enough for fitness differences to produce an evolutionary response. Covariance is a general-purpose measure of association, often used in equations that predict change in a given trait. But a family of statistical measures are relevant in different cases.

A.3. Endler's Summary (2.2)

In Chapter 2 I discussed some simple verbal summaries of evolution by natural selection. Problem cases were used to show their limitations. Here I look at a more careful and detailed summary, due to John Endler (1986: 4).

Natural selection can be defined as a *process* in which:

If a population has:

a. variation among individuals in some attribute or trait: *variation*;

b. a consistent relationship between that trait and mating ability, fertilizing ability, fertility, fecundity, and, or, survivorship: *fitness differences*;

c. a consistent relationship, for that trait, between parents and their offspring, which is at least partially independent of common environmental effects: *inheritance.*

Then:

1. the trait frequency distribution will differ among age classes or life-history stages, beyond that expected from ontogeny;

2. if the population is not at equilibrium, then the trait distribution of all offspring in the population will be predictably different from that of all parents, beyond that expected from conditions *a* and *c* alone.

Conditions *a*, *b*, and *c* are necessary and sufficient for the process of natural selection to occur, and these lead to deductions *1* and *2*. As a result of this process, but not necessarily, the trait distribution may change in a predictable way over many generations.

Endler's formulation takes into account many of the factors which caused problems for the simpler ones. It does not identify fitness with the number of offspring produced by an individual (or the average number produced by a type). A range of properties are associated with fitness in clause *b*, and he is clearly aiming to cover all the features that can affect change in age-structured populations.

However—and largely as a consequence—the formulation has problems. This is because it is expressed as a recipe for change. Some qualifications reduce its predictive content, but that is not what I have in mind. The point is that the ways in which fitness and heredity are handled do not make the formulation applicable as a description of conditions sufficient for change. In clause *b* Endler lists a number of properties related to fitness, but does not collapse these into a single measure. There is no "bottom line" to which survivorship, mating ability, and so on, are said to contribute. If there is no "bottom line," Endler is leaving it open that the mating ability differences might balance out the survival differences, for example, to yield no evolutionary change.

If we leave aside its purported role as a recipe, Endler's formulation is a valuable one. Clause *2*, for example, refers back to the possibility (discussed above) of the inheritance system producing change on its own, and "factors that out" from the change attributed to natural selection. Clause *1* similarly factors out the possible influence of ontogeny. I said in Chapter 2 that there are two ways to approach the abstract description of natural selection. One way is to make idealizations. Then it is possible to keep the summary simple, while also specifying conditions sufficient for change. The other approach is avoid idealization, and try to capture every case, but this "capturing" of the cases no longer involves giving conditions sufficient for change. Endler's formulation, despite being set up like a recipe, does the second.

A.4. Altruism and Correlated Interaction (6.2)

In Chapter 6 two models were compared, represented in Figures 6.1 and 6.2. In the first, a population forms temporary groups at one stage in its life-cycle. Generations are discrete and reproduction is asexual. In the second, the population does not form groups but settles on a lattice.

Models of the evolution of altruism in which group structure is temporary have been extensively discussed (Matessi and Jayakar 1976, Uyenoyama and Feldman 1980, Wilson 1980). The intuitive idea behind the A type being an "altruist" is that all individuals benefit from being in a group containing more, rather than fewer, altruists, but in any given group context, the B type is fitter than A. It is as if the A type "donates" some fitness to everyone in its group. Here is a rule assigning fitnesses in such a case. Let W_i^A be the (absolute) fitness of an individual of the A type in a group with i members of the A type (including itself), and let W_i^B be the fitness of a B individual in a group with i As.

$$W_i^A = z - c + (i - 1)b$$
$$W_i^B = z + ib \tag{A4}$$

Here z is a baseline fitness, c is a cost paid only by A, and b is a benefit received by all individuals from each of the other A-type members of their group. (It is assumed that c and b are both positive.) The outcome of this situation depends not just on the fitnesses but on how groups are formed. If they are formed randomly, the A type is lost, regardless of the details. (Here, and below in this section, a large population is assumed.) But A can prevail (can invade B and remain stable) if groups are formed in a way that "clumps" the two types, so like tends to interact with like. Then the benefits of having As around tend to fall mainly on other As. One index of this clumping is Q:

$$Q = \frac{\sigma^2 - \sigma_R^2}{\sigma_R^2} \tag{A5}$$

Here σ^2 is the variance in the local frequency of A across groups, and σ_R^2 is the variance that would result from random group formation. Then it can be shown that the A type has higher fitness if and only if condition (A6) holds (Wilson 1980, Kerr and Godfrey-Smith 2002b):

$$Q > c/b \tag{A6}$$

So high degrees of "clumping" help the altruist type. This has an obvious relation to Hamilton's rule, discussed below. I now turn to the more unorthodox model, in which individuals settle into a lattice structure or a similar network without group boundaries, interacting with their neighbors. (I will use the term "network" for all structures of this kind in which discrete groups are absent.) Now W_i^A refers to the fitness of an A type with i *neighbors* of the A type, and likewise for W_i^B. Each individual has n neighbors in total. The formula for B's fitness is the same as in (A4); the formula for A's fitness is slightly different given that i now refers only to neighbors: it is $W_i^A = z - c + ib$. Two other parts of such a model are the *neighborhood distributions* and the *network formation rule*. The neighborhood distributions, $N_i^A(t)$ and $N_i^B(t)$, represent the frequencies with which each type encounters neighborhoods with i members of the A type at a given time, t. If we know each of these distributions at a time, these together with the fitnesses and the frequencies of the types suffice to predict change. Here p is the frequency of the A type at t, and p' is its frequency in the next generation.

$$p' = p \sum_{i=0}^{n} N_i^A(t) W_i^A / \overline{W}$$
$$\overline{W} = p \sum_{i=0}^{n} N_i^A(t) W_i^A + (1 - p) \sum_{i=0}^{n} N_i^B(t) W_i^B \tag{A7}$$

Suppose first that neighbors are distributed on the network randomly. Then it can be shown that with the fitnesses above, the A type will be lost (Godfrey-Smith 2008).

So we turn to non-random network formation rules. Complexity arises from the fact that evolutionary change is a consequence of the fitness structure and neighborhood distributions, but what the causal assumptions in the model give us is the network formation rule, and the relation between the two can be complicated. Things are simplified if we can use what can be called a "two-coins model." As in the random case, we imagine predicting each of an individual's neighbors with coin tosses, but now the coin is different according to whether the focal individual is of type A or B. An A individual's neighbors are each predicted with a coin whose probability of choosing A is p_A; for a B individual the coin's probability of choosing A is p_B. This model cannot be applied exactly to the densely packed lattice in Figure 6.2, because each assignment of an individual to the lattice should be constrained by several others, not just one, but it can be used approximately (for example, by filling each row independently and hence having correlation with respect to horizontal neighbors but not vertical ones).

The two-coins principle is used, in effect, by Hamilton (1975) and Nunney (1985), who borrow the parameter F ($0 \leq F \leq 1$) from treatments of inbreeding, for use as a measure of non-random association. F is used along with p to generate the "experienced" frequency of A neighbors for each of the two types.

$$p_A = p + (1 - p)F$$
$$p_B = p - pF \tag{A8}$$

When applicable, this leads to a simple result when we assume the fitness rules in (A4). The A type has a higher fitness if and only if:

$$Fn > c/b \tag{A9}$$

Though they were arrived at by different roads, the Q parameter used for groups and the F parameter used with networks are doing a very similar job. It is also possible to treat the group-structured model as a special case of the neighbor-structured one; discrete groups are one kind of structure to which the model using neighbor interactions can be applied.

A.5. Hamilton's Rule (6.2)

"Hamilton's rule" in its original form states that an altruistic behavior will be favored if and only if $r > c/b$ (Hamilton 1964). Here c is the cost to the actor, b is the benefit received by someone as a consequence of the action, and r is the *coefficient of relatedness* between the actor and recipient. The value of r for human full siblings is $1/2$, for example, as is r between parent and offspring. The rule was initially taken to make good sense of altruistic behavior directed on biological relatives, but to help little with other kinds of altruism and cooperation. Hamilton himself, however, came to see that the principle could be applied more broadly. "[K]inship should be considered just one way of getting positive regression of genotype in the recipient [of altruistic behaviors], and … it is this positive regression that is vitally necessary for altruism" (1975: 337). In Chapter 6 I discussed Queller's formulation of this idea, and here I outline a simplified version of his model and derivation.

Assume an asexual population whose members interact in pairs. Each individual has a value for phenotype, P, which is equal to one if the individual acts altruistically within its pair, and zero otherwise. Each individual also has a value of P^*, which is the phenotype of the individual's partner (again, one if the partner is an altruist, zero otherwise). Each individual also has a value of G, its genotype, and of G^*, the genotype of its partner. (I am re-using the symbol "G" here, which stood for the germ line parameter in earlier chapters, but I will follow Queller's and standard symbolism here. The second use of "G" is confined to this section and there are no germ/soma uses of "G" in this section.) The values of G can initially be thought of as one and zero, for altruistic and selfish respectively, but this assumption does not matter, as we will see below. The cost of being an altruist is c, and the benefit received from having an altruistic partner is b. W_0 is an initial or baseline fitness. Individual i's total fitness can then be written as follows:

$$W_i = W_0 - cP_i + bP_i^* \qquad (A10)$$

Assuming that G is faithfully transmitted from parent to offspring, a Price equation for change in the mean value of G can be written as $\Delta \overline{G} = Cov(W, G)/\overline{W}$. Substituting the right-hand side of (A10) for W and rearranging, the model yields two equivalent criteria for when the mean of G will increase. One of the two is as follows:

$$\frac{Cov(G^*, P)}{Cov(G, P)} > \frac{c}{b} \qquad (A11)$$

The other formulation has $Cov(G, P^*)$ as the left-hand side numerator instead. Either way, "relatedness" is replaced here by an abstract measure of correlation between the phenotypes of those acting and the genotypes of those the actions affect. The recipient need not have the same phenotype as the actor, and the actor need not have the same genotype as the recipient. Further, talk of "genotype" is actually inessential here, as G in the model functions as a quantitative characteristic that is potentially correlated with P and that is passed on in reproduction—those are the *only* constraints on G. Transmission could be cultural, for example, and, more generally, the model does not require that the population can be sorted into discrete types, altruist versus selfish. P could take many values as well. If a population had individuals with many degrees of altruism (as with the case of height), the model would allow us *either* to group them coarsely into the altruist versus selfish (like the tall versus short), *or* to track all the fine differences. The model is thus compatible with evolutionary nominalism of the kind defended above. As discussed earlier, the model can also be extended to cover cases where cooperation is favored through reciprocity (Fletcher and Zwick 2006). If an individual's behavior is sensitive to its circumstances rather than fixed, and cooperation is produced in a discriminate way (perhaps via a "Tit-for-Tat" rule), then $Cov(G^*, P)$ can be high even if pairs initially come together at random.

I will make one additional argument using the model, linking Chapters 2, 6, and 7. The genetic description of evolution is the firmest home of "types" in evolutionary thinking. But evolutionary nominalism applies here as well; genetic types, once we have a DNA sequence of appreciable length, are a coarse-graining just as phenotypic ones are. If we look at many "identical" copies of a gene, we will eventually find a shading-off. Is this genetic sequence a token of the same type as that one if they differ by a silent substitution? Perhaps those do not count, but what about a mutation that affects only an unimportant part of the protein? Genetic sequences are related by distances in a space of substitutions, as well as by type-identity.

This affects discussions of genetic cooperation. The "cooperation" of two identical alleles in closely related cells or organisms is often taken to be readily explicable; they are not really two different things, in the evolutionarily important sense, but instances of a common type. It is the type—the "strategic gene" (Haig 1997)—that does well or badly, and its material tokens rightly behave indifferently between favoring their own copying and favoring their type-mate's copying. But if one strand of DNA acquires a silent substitution, it is not suddenly outside the cooperative fold. Here the Queller formulation of Hamilton's rule is useful. The model (partly via its Price-equation roots) explains donations of fitness between entities that are treated, in the explanation, as unique particulars that can be related by similarity and need not share their type. The case of two gene copies with identical sequence is treated as an extreme case of a more general phenomenon.

A.6. Connection, Modification, and Descent

Evolution in a Darwinian population is one kind of change in a system of objects over time, and a focus of this book has been the idea that Darwinian evolution shades into other kinds of change. A connected topic is the relation between different "levels of description." Here I have in mind not just levels of *selection*, as in Chapter 6, but the fact that Darwinian populations are physical systems, and at the physical level different kinds of description apply to them. Each Darwinian individual is a collection of physical particles, moving through space and time, constantly losing and gaining matter. Darwinian processes become visible via a "zooming-out" from a mass of physical events. In this section I present a formal way of representing and investigating some of these issues. All the work in this section was done in collaboration with Ben Kerr, and the equation (A12) was proved by him.

Suppose a system consists of two collections of things existing at different times, with at least some causal connections linking the entities present at different times. The two times will be labeled t^a and t_d, for the "ancestral" and "descendant" time points, respectively, and the collections of entities will be also be referred to as ancestral and descendant. Throughout, superscripts will indicate ancestral properties and subscripts will indicate those of descendants.

Assume that at least some of the descendant entities are connected to some of the ancestral entities. This "connection" can be thought of initially as some sort of causal responsibility, but that term is understood very broadly. If an object

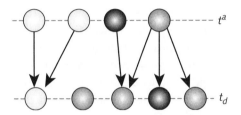

Figure A.1: Ancestral and descendant entities.

persists intact from t^a to t_d, that is sufficient for connection. Familar kinds of reproduction also count. But *any* pattern of connectivity is allowed in the analysis. As Figure A.1 shows, the ancestors may differ in the number of connections they have to members of the descendant ensemble, and the descendants can also differ with respect to the number of connections they have to the ancestrals. There can be ancestors with no descendants, and descendants with no ancestors. The only constraint is that at least one connection exists.

Below I will describe a representation of change in systems of this kind. First, though, it is useful to take a step back from what is assumed so far. Imagine that we are at an earlier stage of analysis of the system in Figure A.1. We have not yet recognized distinct objects making up the ancestral and descendant ensembles. All we know is that the entire system at t^a gives rise to entire system at t_d. To reach the stage of analysis pictured in Figure A.1 we have to first recognize separate objects, at both time points, and secondly limit the connections recognized between them. These preliminary stages are represented in Figure A.2, (a) and (b).

The second move, from the representation in Figure A.2(b) to Figure A.1, involves a kind of coarse-graining. We can assume everything present at t^a has *some* effect on everything at t_d—there are minute gravitational effects, if nothing else. To reach a picture with limited connectivity we ignore many of these influences, and treat only some as significant. The earlier move, from Figure A.2(a) to A.2(b) is more philosophically controversial, but may also involve a similar kind of coarse-graining. To reach A.2(b) we treat some parts of the overall system as partially independent of the others, with an identity that is portable

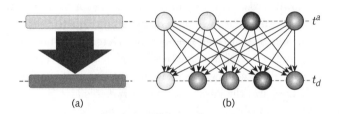

Figure A.2: Stages of analysis preliminary to Figure A.1.

across changes in other members of the collection. This is related to the distinction made in Chapter 8 between populations and highly integrated networks.

Suppose that we have reached the kind of picture seen in Figure A.1, with collections of distinct entities and limited connections between them. The next aim is to represent change over the time interval. Let there be n^a entities at t^a and n_d entities at t_d. Let C_j^i be an indicator variable for connection between ancestral entity i and descendant entity j. So:

$$C_j^i = \begin{cases} 1 \text{ if ancestral entity } i \text{ connects to descendant entity } j \\ 0 \text{ if ancestral entity } i \text{ does not connect to descendant entity } j \end{cases}$$

Thus, ancestral entity i connects to a total of $C_*^i = \sum_{j=1}^{n_d} C_j^i$ descendant entities, and descendant entity j connects to $C_j^* = \sum_{i=1}^{n^a} C_j^i$ ancestral entities. These are absolute measures of connectedness for ancestors and descendants. We can also define two relative measures of connectedness, $\tilde{C}_*^i$ and $\tilde{C}_j^*$, by dividing C_*^i and C_j^* by C_*^* / n^a and C_*^* / n_d respectively. That is, we divide the two absolute measures by the average connectedness of ancestors (in one case) and of descendants (in the other). Here C_*^* is the total number of connections, or $\sum_{i=1}^{n^a} \sum_{j=1}^{n_d} C_j^i$.

X is some measurable characteristic of the entities. Let the value of X for ancestral entity i be X^i and that of descendant entity j be X_j. The mean character values in the ancestral and descendant ensembles are $\overline{X}^a = \frac{1}{n^a} \sum_{i=1}^{n^a} X^i$ and $\overline{X}_d = \frac{1}{n_d} \sum_{j=1}^{n_d} X_j$. Change can then be represented with an equation linking these two means; let $\Delta \overline{X}$ be the difference between the means, or $\overline{X}_d - \overline{X}^a$. It can be shown that:

$$\Delta \overline{X} = Cov(\tilde{C}_*^i, X^i) + E(\Delta X_j^i) - Cov(\tilde{C}_j^*, X_j) \tag{A12}$$

Here $\Delta X_j^i = C_j^i(X_j - X^i)$, the change in character across a particular connection; $E(\Delta X_j^i)$ is the average change across a connection.

Despite the complicated set-up, this equation is easy to interpret (see Kerr and Godfrey-Smith, forthcoming, for more detail). The first two terms in the right-hand side map to the terms found in a standard Price equation. The first term is a covariance between the character value of each ancestor and the number of descendants to which it is connected, relativized to the overall degree of connectedness seen in ancestors. $\tilde{C}_*^i$ is thus a kind of fitness measure, treating the presence of a downwards arrow as a unit of influence for that ancestral entity. So $Cov(\tilde{C}_*^i, X^i)$ measures the covariance of ancestral character with fitness. The second term measures the overall tendency of divergence to take place over a connection—it is like a "transmission bias" term. The third term, which is not part of a standard Price equation, is like a mirror-image of the first term, the fitness term. It measures the covariance between *descendant* character and the number of ancestors to which the descendant is connected, relative to the overall degree of connectedness seen in descendants.

The Price equation is often seen as giving a complete decomposition of evolutionary change. But change is consistent with zero values for the two standard Pricean terms. The explanation of the "missing term" is as follows. It is usually assumed that the members of a parental generation may differ with respect to their number of offspring, but it is *not* usually assumed that the members of the offspring generation might differ with respect to the number of their *parents*. The present model, in contrast, makes no prior assumptions regarding the number of parents that an individual has; any pattern of connectivity is treated as possible, including one-to-many and many-to-one connections in each direction. The standard Price equation covers a special case that arises via an (often reasonable) simplifying assumption about the pattern of connectivity between the ancestral and descendant ensembles.

One simple example to illustrate the role of the third term is migration into the population from outside. A migrant is, in the context of this analysis, a descendant without an ancestor. When some individuals in the descendant ensemble are migrants and some are not, and the migrants differ in character from the locals, $Cov(\tilde{C}_j^*, X_j)$ will be non-zero. Another example is a mixture of sexual and asexual reproduction (as seen in Figure A.1). Then, again, individuals will differ in their number of parents, and if those with more or fewer parents differ in character from the others, $Cov(\tilde{C}_j^*, X_j)$ will be non-zero. It has often been noted that the structure of mainstream evolutionary theory is better designed for fruit flies and birds than it is for plants and for animals which show mixtures of sex and asexuality (see Chapter 4, along with Jackson et al. 1985, Tuomi and Vuorisalo 1989b). The generalization of the Price equation here equips it to deal with those cases, without reducing them (as may also be done) to a genetic level at which reproduction is more uniform. In contrast with the usual Price equation, the analysis here is also reversible. Equation (A12) treats change from the ancestral to descendant ensembles as a consequence of ancestral fitness differences, transmission bias, and differences in a descendant-focused mirror-image of fitness. But as any pattern of connectivity is allowed, an analysis using (A12) could describe change *from* an "descendant" ensemble *to* an "ancestral."

The utility of the Price equation derives in part from the way it can be applied to hierarchically structured systems. Price's second or "expectation" term, which corresponds to the second term in (A12), can be decomposed into a lower-level covariance term and a lower-level expectation. Equation (A12) also has this feature, but the expectation term breaks down into *three* lower-level terms, each corresponding to the terms described above (see Kerr and Godfrey-Smith, forthcoming).

When these points about hierarchy are made, it is usually assumed that the analyst knows in advance that there is a lower level of reproducing entities. The present framework can be used to represent how such conclusions may be reached. To see this, take a step back to Figure A.2. The equation above can be applied to the cases in Figure A.2, but not in an informative way. The analysis will be trivial in Figure A.2(a), as there is only one connection and one member of each ensemble. So the change over that one connection is $\Delta \overline{X}$. In the case of Figure A.2(b) the analysis will also be relatively uninformative, though not trivial. In Figure A.2(b), all the ancestors are jointly responsible for all the descendants, and only the second term can be non-zero. The breakdown given by the equation becomes informative when a significant role is being played by differing descendant number (differential fitness), differing ancestor number, or both. Change has a more Darwinian character when a significant role is played by the first term, the differential fitness term. It has a more "transformational" character (Lewontin 1983) when much of the weight is carried by some regular principle of change over a connection described in the second term. There is not an existing label that captures change due mainly to the third term, which is a matter of "differential convergence."

Returning to questions about hierarchy: when someone wonders whether a hierarchical analysis will be informative, they begin, in effect, by treating the entities that make up our initial "focal" level (the circular shapes in Figure A.1) as if they were *each* like the shapes in Figure A.2(a)—undifferentiated wholes linked by single connections across the time interval. They may then ask whether these entities can be treated as collectives—whether they can be broken down into smaller units that enter into ancestor/descendant relations of their own. This question can be addressed using the same criteria described above for the initial or "focal" level. Is there a natural division of the focal entities into parts at all? If there is, does this division allow us to recognize a reasonably sparse, and hence informative, pattern of connection between sub-entities across the time interval? This is pictured in Figure A.3. Here I suppose that two higher-level entities are connected if and only if there is at least sub-entity connection between them. Alternatively, working from the focal level "up," we can assess whether the focal entities can be collected together into larger units, revealing a Darwinian pattern at a higher level.

The analysis is obviously very general with respect to the entities making up the population. They might be organisms, time-slices of organisms, cells, genes, groups, or cultural variants. It is not assumed that reproduction is synchronized; the descendant entities could be at an earlier stage of individual development than the ancestors, and the two ensembles might differ internally in the same way. There may be intervening generations not represented. In addition, no distinction is made between persisting across the time interval and asexual reproduction accompanied by death of the parent. Either way, a single ancestor gives rise to a single descendant. In Chapter 5 much was made of the distinction between reproducing and persisting. The features of reproduction captured by

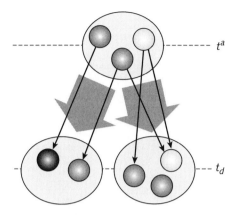

Figure A.3: Connectivity at two levels.

parameters B and G of that chapter are not automatically given a role in this analysis. That is a way in which the framework here is incomplete.

In Section A.1 I distinguished two kinds of analyses of evolutionary change: those that apply over many time-steps ("dynamically sufficient" representations, or "recursions"), and "single-step" analyses that only contain information bearing on one interval or generation. Using this model we can try to describe the kinds of features that will make it possible to describe a system with a recursive expression. Two kinds of simplicity should rise to this situation. First, there may be a simple rule for change over a connection. Second, there may be a simple rule relating the character of an ancestor (X^i) to the number of connections it has to the descendant generation. The result may then be a compact dynamic rule, applicable over many time-steps, such as a discrete replicator dynamics with mutation (Nowak 2006). That dynamic requires that each descendant has only one ancestor, and change over the connection is described with a fixed high probability of faithful transmission and a small probability of change to a different state. It also requires that the fitness of an ancestor either has a fixed association with its character, or is a function of factors that can themselves be predicted evolutionarily, such as the frequency of a type. Then we have fitnesses systematically associated with repeatable types.

There will be other simple rules possible as well, which do not require that each descendant has only one ancestor. But evolutionary processes will only be orderly and tractable over long intervals when there is reasonable simplicity in the rules that determine, as a function of ancestral properties, which descent lines or connections are going to arise, and what sort of population will arise from those connections. This gives us a way of thinking about the contrast between the orderly processes of Mendelian inheritance and the less orderly processes of cultural change.

BIBLIOGRAPHY

Abrams, J. and Eickwort, G. C. (1981). "Nest Switching and Guarding by the Communal Sweat Bee *Agapostemon virescens* (Hymenoptera, Halictidae)." *Insectes Sociaux* 28: 105–16.

Adelson, G. (forthcoming). "Naming Evolutionary Patterns and Processes: Continuing the Campaign Against Essentialism."

Amundson, R. (1989). "The Trials and Tribulations of Selectionist Explanations," in K. Hahlweg and C. A. Hooker (ed.), *Issues in Evolutionary Epistemology*. Albany, NY: SUNY Press, 413–32.

Anderson, C. and McShea, D. W. (2001). "Individual versus Social Complexity, with Particular Reference to Ant Colonies." *Biological Review* 76: 211–37.

Ariew, A. (2008). "Population Thinking," in M. Ruse (ed.), *Handbook of Philosophy of Biology*. Oxford: Oxford University Press, 64–86.

—— and Lewontin, R. C. (2004). "The Confusions of Fitness." *British Journal for the Philosophy of Science* 55: 347–63.

Avital, E. and Jablonka, E. (2000). *Animal Traditions: Behavioral Inheritance in Evolution*. Cambridge: Cambridge University Press.

Axelrod, R. and Hamilton, W. D. (1981). "The Evolution of Cooperation." *Science* 211: 1390–6.

Bateson, P. (1978). "Book Review: *The Selfish Gene* by Richard Dawkins." *Animal Behaviour* 26: 316–18.

—— (2006). "The Nest's Tale. A reply to Richard Dawkins." *Biology and Philosophy* 21: 553–8.

Beatty, J. (1984). "Chance and Natural Selection." *Philosophy of Science* 51: 183–211.

—— and Finsen, S. (1989). "Rethinking the Propensity Interpretation: A Peek Inside Pandora's Box," in M Ruse (ed.), *What the Philosophy of Biology is: Essays for David Hull*. Dordecht: Kluwer, 17–31.

Benirshke, K., Anderson, J. M., and E., Brownhill L. (1962). "Marrow Chimerism in Marmosets," *Science* 138: 513–15.

Bergman, A. and Feldman, M. W. (1995). "On the Evolution of Learning: Representation of a Stochastic Environment." *Theoretical Population Biology* 48: 251–76.

Bishop, C. D., Erezyilmaz, D. F., Flatt, T., et al. (2006). "What is Metamorphosis?" *Integrative and Comparative Biology* 46: 655–61.

Blute, M. (2007). "The Evolution of Replication." *Biological Theory* 2: 10–22.

Bonner, J. T. (1959). *The Cellular Slime Molds*. Princeton, NJ: Princeton University Press.

———— (1974). *On Development: The Biology of Form*. Cambridge, MA: Harvard University Press.

Boss, P. K. and Thomas, M. R. (2002). "Association of Dwarfism and Floral Induction with a Grape 'Green Revolution' Mutation." *Nature* 416: 847–50.

Bouchard, F. (forthcoming). "Causal Processes, Fitness and the Differential Persistence of Lineages." *Philosophy of Science*, PSA 2006 Proceedings.

Bourke, A. F. (1888). "Worker Reproduction in the Higher Eusocial Hymenoptera." *Quarterly Review of Biology* 63: 291–311.

Bowles, S. and Gintis, H. (2003). "The Origins of Human Cooperation," in P. Hammerstein (ed.), *The Genetic and Cultural Origins of Cooperation*. Cambridge, MA: MIT Press, 429–44.

Boyd, R. and Richerson, P. J. (1985). *Culture and the Evolutionary Process*. Chicago, IL: University of Chicago Press.

Brandon, R. N. (1978). "Adaptation and Evolutionary Theory." *Studies in History and Philosophy of Science* 9: 181–206.

———— (1988). "The Levels of Selection: A Hierarchy of Interactors," in H. C. Plotkin (ed.), *The Role of Behavior in Evolution*. Cambridge, MA: MIT Press, 51–72.

———— (unpublished). "Inheritance Biases and the Insufficiency of Darwin's Three Conditions."

Buller, D. J. (ed.), (1999). *Function, Selection, and Design*. Albany, NY: SUNY Press.

Burian, R. M. (1992). "Adaptation: Historical Perspectives," in E. A. Lloyd and E. Fox Keller (eds.), *Keywords in Evolutionary Biology*. Cambridge, MA: Harvard University Press, 7–12.

Burt, A. and Trivers, R. (2006). *Genes in Conflict: The Biology of Selfish Genetic Elements*. Cambridge, MA: Harvard University Press.

Buss, L. W. (1987). *The Evolution of Individuality*. Princeton, NJ: Princeton University Press.

Calcott, B. (2008). "The Other Cooperation Problem: Generating Benefit," *Biology and Philosophy* 23: 179–203.

Campbell, D. T. (1974). "Evolutionary Epistemology," in P. A. Schillp (ed.), *The Philosophy of Karl Popper*. La Salle, IL: Open Court, 413–63.

Cavalli-Sforza, L. L. and Feldman, M. W. (1981). *Cultural Transmission and Evolution: A Quantitative Approach*. Princeton, NJ: Princeton University Press.

Charlesworth, B. (1994). *Evolution in Age-Structured Populations* (2nd edn.). Cambridge: Cambridge University Press.

_____ and Giesel, J. T. (1972). "Selection in Populations with Overlapping Generations II: Relations between Gene Frequency and Demographic Variables." *American Naturalist* 106: 388–401.

Cook, R. E. (1980). "Reproduction by Duplication." *Natural History* 89: 88–93.

Cosmides, L. and Tooby, J. (1992). "Cognitive Adaptations for Social Exchange," in J. Barkow, L. Cosmides, and J. Tooby (eds.), *The Adapted Mind: Evolutionary Psychology and the Generation of Culture.* New York: Oxford University Press, 163–228.

Crespi, B. J. and Yanega, D. (1995). "The Definition of Eusociality." *Behavioural Ecology,* 6, 109–15.

Crow, J. F. (1986). *Basic Concepts in Population, Quantitative, and Evolutionary Genetics.* New York, NY: W. H. Freeman.

Damuth, J. and Heisler, I. L. (1988). "Alternative Formulations of Multilevel Selection." *Biology and Philosophy* 3: 407–30.

Darden, L. and Cain, J. A. (1989). "Selection Type Theories." *Philosophy of Science* 56: 106–29.

Darwin, C. (1859/1964). *On the Origin of Species* (Facsimile of the 1st edn.). Cambridge, MA: Harvard University Press.

Dawkins, R. (1976). *The Selfish Gene.* Oxford: Oxford University Press.

_____ (1978). "Replicator Selection and the Extended Phenotype," reprinted in E. Sober (ed.), *Conceptual Issues in Evolutionary Biology.* Cambridge, MA: MIT Press, 1984, 125–41.

_____ (1982a). *The Extended Phenotype: The Gene as the Unit of Selection.* Oxford: W. H. Freeman.

_____ (1982b). "Replicators and Vehicles," in King's College Sociobiology Group (ed.), *Current Problems in Sociobiology.* Cambridge: Cambridge University Press, 45–64.

_____ (1986). *The Blind Watchmaker.* New York, NY: Norton.

_____ (1989). "The Evolution of Evolvability," in C. Langton (ed.), *Artificial Life VI.* California: Addison-Wesley, 201–20.

_____ (2006). *The God Delusion.* Boston, MA: Houghton Mifflin.

Day, T. and Otto, S. P. (2001). "Fitness." *Encyclopaedia of Life Sciences*: Nature Publishing Group. <http://www.els.net>

Dempster, E. R. (1955). "Maintenance of Genetic Heterogeneity." *Cold Spring Harbor Symposia on Quantitative Biology* 20: 25–32.

Dennett, D. C. (1974). "Why the Law of Effect Won't Go Away," reprinted in *Brainstorms: Philosophical Essays on Mind and Cognition.* Cambridge, MA: Bradford Books/MIT Press, 71–89.

_____ (1991). *Consciousness Explained.* Boston, MA: Little, Brown.

_____ (1995). *Darwin's Dangerous Idea: Evolution and the Meanings of Life.* New York, NY: Simon & Schuster.

Dennett, D. C. (2001). "The Evolution of Evaluators," in A. Nicita and U. Pagano (eds.), *The Evolution of Economic Diversity*. London: Routledge, 66–81.

—— (2006). *Breaking the Spell: Religion as a Natural Phenomenon*. New York, NY: Viking.

Downes, S. M. (2007). "Heredity and Heritability," *Stanford Encyclopedia of Philosophy (Fall 2007 Edition)* <http://plato.stanford.edu/archives/fall2007/entries/heredity/>.

Drake, J. W., Charlesworth, B., Charlesworth, D., and Crow, J. F. (1998). "Rates of Spontaneous Mutation." *Genetics* 148: 1667–86.

Duffy, J. E. (1996). "Eusociality in a Coral Reef Shrimp." *Nature* 381: 512–14.

Dugatkin, L. A. (2002). "Cooperation in Animals: An Evolutionary Overview." *Biology and Philosophy* 17: 459–76.

Dumais, J. and Kwiatkowska, D. (2001). "Analysis of Surface Growth in Shoot Apices." *The Plant Journal* 31: 229–41.

Dupré, J. (2002), *Humans and Other Animals*. Oxford: Oxford University Press.

Durham, W. H. (1991). *Coevolution: Genes, Culture and Human Diversity*. Stanford, CA: Stanford University Press.

Eigen, M. and Schuster, P. (1979). *The Hypercycle: A Principle of Natural Self-Organization*. Berlin: Springer-Verlag.

Eldredge, N. (1985). *Unfinished Synthesis: Biological Hierarchies and Modern Evolutionary Thought*. Oxford: Oxford University Press.

Endler, J. A. (1986). *Natural Selection in the Wild*. Princeton, NJ: Princeton University Press.

Eshel, I. and Cavalli-Sforza, L. (1982). "Assortment of Encounters and Evolution of Cooperativeness." *Proceedings of the National Academy of Sciences (USA)* 79: 1331–15.

Fisher, R. A. (1930). *The Genetical Theory of Natural Selection*. Oxford: Clarendon Press.

Fletcher, J. A. and Zwick, M. (2006). "Unifying the Theories of Inclusive Fitness and Reciprocal Altruism." *American Naturalist* 168: 252–62.

Forber, P. (2005). "On the Explanatory Roles of Natural Selection." *Biology and Philosophy* 20: 329–42.

Fracchia, J. and Lewontin, R. C. (1999). "Does Culture Evolve?" *History and Theory* 38: 52–78.

Francis, R. C. (2004). *Why Men Won't Ask for Directions: The Seductions of Sociobiology*. Princeton, NJ: Princeton University Press.

Frank, S. A. (1995). "George Price's Contributions to Evolutionary Genetics." *Journal of Theoretical Biology* 175: 373–88.

—— (1998). *Foundations of Social Evolution*. Princeton, NJ: Princeton University Press.

—— (2007). *Dynamics of Cancer: Incidence, Inheritance, and Evolution*. Princeton, NJ: Princeton University Press.

_____ and Slatkin, M. (1990). "Evolution in a Variable Environment." *American Naturalist* 136: 244–60.

Franklin, L. R. (2007). "Bacteria, Sex, and Systematics." *Philosophy of Science* 74: 69–95.

Franks, T., Botta, R., Thomas, M. R., and Franks, J. (2002). "Chimerism in Grapevines: Implications for Cultivar Identity, Ancestry and Genetic Improvement." *Theoretical and Applied Genetics* 104: 192–9.

Froissart R., Roze D., Uzest M., Galibert L., Blanc S., and Michalakis, Y. (2005). "Recombination Every Day: Abundant Recombination in a Virus during a Single Multi-Cellular Host Infection." *Public Library of Science: Biology* 3: e89.

Frumkin, D., Wasserstrom, A., Kaplan, S., Feige, U., and Shapiro, E. (2005). "Genomic Variability within an Organism Exposes Its Cell Lineage Tree." *Public Library of Science: Computational Biology* 1: 382–94.

Futuyma, D. J. (1986). *Evolutionary Biology* (2nd edn.) Sunderland, MA: Sinauer.

Gannett, L. (2003). "Making Populations: Bounding Genes in Space and in Time." *Philosophy of Science* 70: 989–1001.

Gavrilets, S. (2004). *Fitness Landscapes and the Origin of Species.* Princeton, NJ: Princeton University Press.

Ghiselin, M. T. (1974). "A Radical Solution to the Species Problem." *Systematic Zoology* 23: 536–44.

Giere, R. N. (1988). *Explaining Science: A Cognitive Approach.* Chicago, IL: University of Chicago Press.

Gill, D. E., Chao, L., Perkins, S. L., and Wolf, J. B. (1995). "Genetic Mosaicism in Plants and Clonal Animals." *Annual Review of Ecology and Systematics* 26: 423–44.

Gillespie, J. H. (1972). "The Effects of Stochastic Environments on Allele Frequencies in Natural Populations." *Theoretical Population Biology* 3: 241–8.

Godfrey-Smith, P. (1996). *Complexity and the Function of Mind in Nature.* Cambridge: Cambridge University Press.

_____ (2000). "The Replicator in Retrospect." *Biology and Philosophy* 15: 403–23.

_____ (2006). "The Strategy of Model-Based Science." *Biology and Philosophy* 21: 725–40.

_____ (2007a), "Conditions for Evolution by Natural Selection." *The Journal of Philosophy* 104: 489–516.

_____ (2007b). "Information in Biology," in D. L. Hull and M. Ruse (eds.), *The Cambridge Companion to the Philosophy of Biology.* New York, NY: Cambridge University Press, 103–19.

_____ (2008). "Varieties of Population Structure and the Levels of Selection." *British Journal for the Philosophy of Science* 59: 25–50.

Gould, S. J. (1976). "Darwin's Untimely Burial." *Natural History* 85: 24–30.

_____ (1985). "A Most Ingenious Paradox," in *The Flamingo's Smile.* New York, NY: Norton, 78–98.

Gould, S. J. (2002). *The Structure of Evolutionary Theory*. Cambridge, MA: Harvard University Press.

Gould, S. J. and Eldredge, N. (1977). "Punctuated Equilibria: The Tempo and Mode of Evolution Reconsidered." *Paleobiology* 3: 115–51.

Grafen, A. (1985). "A Geometric View of Relatedness." *Oxford Surveys in Evolutionary Biology* 2: 28–90.

—— (2007). "The Formal Darwinism Project: A Mid-Term Report." *Journal of Evolutionary Biology* 20: 1243–54.

Gray, R. D. and Atkinson, Q. D. (2003). "Language-Tree Divergence Times Support the Anatolian Theory of Indo-European Origin." *Nature* 426: 435–39.

—— Greenhill, S. J., and Ross, R. M. (forthcoming). "The Pleasures and Perils of Darwinizing Culture (with Phylogenies)." *Biological Theory*.

Griesemer, J. (2000). "The Units of Evolutionary Transition." *Selection* 1: 67–80.

—— (2005). "The Informational Gene and the Substantial Body: On the Generalization of Evolutionary Theory by Abstraction," in M. Jones and N. Cartwright (eds.), *Idealization XII: Correcting the Model, Idealization and Abstraction in the Sciences*. Amsterdam: Rodopi, 59–115.

—— and Wade, M. (2000). "Populational Heritability: Extending Punnett Square Concepts to Evolution at the Metapopulation Level." *Biology and Philosophy* 15: 1–17.

Griffiths, P. E. (2002). "What is Innateness?" *The Monist* 85: 70–85.

—— and Gray, R. D. (1994). "Developmental Systems and Evolutionary Explanation." *The Journal of Philosophy* 91: 277–304.

—— and M. Neumann-Held (1999). "The Many Faces of the Gene." *BioScience* 49: 656–62.

—— and Stotz, K. (2006). "Genes in the Postgenomic Era?" *Theoretical Medicine and Bioethics* 27: 499–521.

Grimes, G. (1982). "Nongenic Inheritance: A Determinant of Cellular Architecture." *BioScience* 32: 279–80.

Grosberg, R. and Strathman, R. (1998). "One Cell, Two Cell, Red Cell, Blue Cell: The Persistence of a Unicellular Stage in Multicellular Life Histories." *Trends in Ecology and Evolution* 13: 112–16.

Haag-Liautard, C., Dorris, M., Maside, X., Macaskill, S., Halligan, D. L., Charlesworth, B., and Keightley, P. D. (2007). "Direct Estimation of Per Nucleotide and Genomic Deleterious Mutation Rates in *Drosophila*." *Nature* 445: 82–5.

Haber, M. H. and Hamilton, A. (2005). "Coherence, Consistency, and Cohesion: Clade Selection in Okasha and Beyond." *Philosophy of Science* 72: 1026–40.

Haig, D. (1997). "The Social Gene," in J. R. Krebs and N. B. Davies (eds.), *Behavioural Ecology* (4th edn.). Cambridge, MA: Blackwell Publishing, 284–304.

—— (1999), "What is a Marmoset?" *American Journal of Primatology,* 49, 285–96.

_____ and Grafen, A. (1991). "Genetic Scrambling as a Defence Against Meiotic Drive." *Journal of Theoretical Biology* 153: 531–58.

Haldane, J. B. S. (1996). "The Negative Heritability of Neonatal Jaundice." *Annals of Human Genetics* 60: 3–5.

Hamilton, W. D. (1964). "The Genetical Evolution of Social Behaviour, I." *Journal of Theoretical Biology* 7: 1–16.

_____ (1975). "Innate Social Aptitudes of Man: An Approach from Evolutionary Genetics." Reprinted in *Narrow Roads of Gene Land, Vol 1.* Oxford: W. H. Freeman, 1997, 133–53.

Hardin, G. (1968). "The Tragedy of the Commons." *Science* 162: 1243–8.

Harper, J. L. (1977). *Population Biology of Plants.* London: Academic Press.

_____ and Bell, A. D. (1979), "The Population Dynamics of Growth Form in Organisms with Modular Construction," in R. M. Anderson, B. D. Turner, and L. R. Taylor (eds.), *Population Dynamics: The 20th Symposium of the British Ecological Society.* Oxford: Blackwell, 29–52.

Henrich, J. and Boyd, R. (1998). "The Evolution of Conformist Transmission and the Emergence of Between-Group Differences." *Evolution and Human Behavior* 19: 215–42.

_____ _____ (2002). "On Modeling Cognition and Culture: Why Cultural Evolution does not Require Replication of Representations." *Journal of Cognition and Culture* 2: 87–112.

Herron, M. and Michod, R. (2008). "Evolution of Complexity in the Volcocine Algae: Transitions in Individuality through Darwin's Eye." *Evolution* 62: 436–51.

Heywood, J. S. (2005). "An Exact Form of the Breeder's Equation for the Evolution of a Quantitative Trait under Natural Selection." *Evolution* 59: 2287–98.

Hocquigny, S., Pelsy, F., Dumas, V., Kindt, S., Heloir, M-C., and Merdinoglu, D. (2004). "Diversification within Grapevine Cultivars goes through Chimeric States." *Genome* 47: 579–89.

Hodge, M. J. S. (1987). "Natural Selection as a Causal, Empirical, and Probabilistic Theory," in L. Krüger (ed.), *The Probabilistic Revolution.* Cambridge, MA: MIT Press, 233–70.

Hodgson, G. and Knudson, T. (2006). "The Nature and Units of Social Selection." *Journal of Evolutionary Economics* 16: 477–89.

Hollick J., Dorweiler J., and Chandler V., (1997). "Paramutation and Related Allelic Interactions." *Trends in Genetics* 13: 302–8.

Hull, D. (1978). "A Matter of Individuality." *Philosophy of Science* 45: 335–60.

_____ (1980). "Individuality and Selection." *Annual Review of Ecology and Systematics* 11: 311–32.

_____ (1988). *Science as a Process: An Evolutionary Account of the Social and Conceptual Development of Science.* Chicago, IL: University of Chicago Press.

Hull, D., Langman, R. and Glenn, S., (2001). "A General Analysis of Selection."
Behavioral and Brain Sciences 24: 511–73.

Hutchings, M. J. and Booth, D. (2004). "Much Ado About Nothing... So Far?"
Journal of Evolutionary Biology 17: 1184–6.

Huxley, T. H. (1852). "Upon Animal Individuality." *Proceedings of the Royal
Institution of Great Britain* 1: 184–9.

Jablonka, E. and Lamb, M. J. (1995). *Epigenetic Inheritance and Evolution: The
Lamarckian Dimension*. Oxford and New York: Oxford University Press.

——— ——— (2005). *Evolution in Four Dimensions*. Cambridge, MA: MIT Press.

Jackson, J. B. C. (1985). "Distribution and Ecology of Clonal and Aclonal Benthic
Invertebtrates," in Jackson, Buss, and Cook (1985), 297–356.

——— and Coates, A. G. (1986). "Life Cycles and Evolution of Clonal (Modular)
Animals." *Philosophical Transactions of the Royal Society of London. Series B,
Biological Sciences* 313: 7–22.

——— Buss, L., and R. Cook, (eds.) (1985). *Population Biology and Evolution of
Clonal Organisms*. New Haven: Yale University Press.

Jacquard, A. (1983). "Heritability: One Word, Three Concepts." *Biometrics* 39:
465–77.

Janzen, D. H. (1977). "What are Dandelions and Aphids?" *American Naturalist*
111: 586–9.

Jenkin, F. (1867). "*The Origin of Species*." *The North British Review* 46 (92):
151–171.

Keller, L. (ed.) (1999). *Levels of Selection in Evolution*. Princeton, NJ: Princeton
University Press.

Kerr, B. (2007). "Niche Construction and Cognitive Evolution." *Biological Theory*
2: 250–62.

——— and Godfrey-Smith, P. (2002a). "Individualist and Multi-Level Perspectives on Selection in Structured Populations." *Biology and Philosophy* 17:
477–517.

——— ——— (2002b). "On Price's Equation and Average Fitness." *Biology and
Philosophy* 17: 551–65.

——— ——— (forthcoming). "Generalization of the Price Equation for Evolutionary
Change." To appear in *Evolution*.

——— Neuhauser, C, Bohannan, B., and Dean, A. (2006). "Local Migration Pro-
motes Competitive Restraint in a Host-Pathogen 'Tragedy of the Commons'."
Nature 442: 75–8.

Kirk, D. L. (1998). *Volvox: Molecular-Genetic Origins of Multicellularity and
Cellular Differentiation*. Cambridge and New York: Cambridge University
Press.

——— (2005). "A Twelve-Step Program for Evolving Multicellularity and a Division of Labor." *Bioessays* 27: 299–310.

Kirschner, M. and Gerhart, J. (1998). "Evolvability." *Proceedings of the National Academy of Sciences (USA)* 95: 8420–7

Kitcher, P. S. (1985). *Vaulting Ambition: Sociobiology and the Quest for Human Nature.* Cambridge, MA: MIT Press.

Klekowski, E. J. (1988). *Mutation, Developmental Selection, and Plant Evolution.* New York, NY: Columbia University Press.

Krimbas, C. (2004). "On Fitness." *Biology and Philosophy* 19: 185–203.

Kuhn, T. S. (1962). *The Structure of Scientific Revolutions.* Chicago, IL: University of Chicago Press.

Kukuk, P. F. and Sage, G. K. (1994). "Reproductivity and Relatedness in a Communal Halictine Bee *Lasioglossum (Chilalictus) hemichalceum.*" *Insectes Sociaux* 41: 443–55.

Kutschera, U. and Niklas, K. J. (2005). "Endosymbiosis, Cell Evolution, and Speciation." *Theory in Biosciences* 124: 1–24.

Lane, N. (2005). *Power, Sex, Suicide: Mitochondria and the Meaning of Life.* Oxford: Oxford University Press.

Langton, R. and D. K. Lewis (1998). "Defining 'Intrinsic'." *Philosophy and Phenomenological Research* 58: 333–45.

Lehmann, L. and Keller, L. (2006). "The Evolution of Cooperation and Altruism: A General Framework and a Classification of Models." *Journal of Evolutionary Biology* 19: 1365–76.

Lennox, J. (2006). "Aristotle's Biology." *The Stanford Encyclopedia of Philosophy (Fall 2006 Edition)* <http://plato.stanford.edu/archives/fall2006/entries/aristotle-biology/>.

Levy, A. (forthcoming). "Source-Based Learning and the Evolution of Morality."

Lewens, T. (2007). *Darwin.* London: Routledge.

Lewontin, R. C. (1955). "The Effects of Population Density and Composition on Viability in *Drosophila melanogaster.*" *Evolution* 9: 27–41.

—— (1970). "The Units of Selection." *Annual Review of Ecology and Systematics* 1: 1–18.

—— (1985). "Adaptation," in R. Levins and R. C. Lewontin (eds.), *The Dialectical Biologist.* Cambridge, MA: Harvard University Press, 65–84.

Lloyd, E. (1988). *The Structure and Confirmation of Evolutionary Theory.* New York, NY: Greenwood Press.

—— (2001). "Units and Levels of Selection: An Anatomy of the Units of Selection Debates," in R. Singh, C. Krimbas, D. Paul, and J. Beatty (eds.), *Thinking about Evolution: Historical, Philosophical, and Political Perspectives.* New York, NY: Cambridge University Press, 267–91.

—— and Gould, S. J. (1993). "Species Selection on Variability." *Proceedings of the National Academy of Sciences (USA)* 90: 595–9.

Lombrozo, T., Shtulman, A., and Weisberg, M. (2006). "The Intelligent Design Controversy: Lessons from Psychology and Education." *Trends in Cognitive Sciences* 10: 56–7.

Loux, M. (2002). *Metaphysics: A Contemporary Introduction*. London: Routledge.

Loxdale, H. D. and Lushai, G. (2003). "Rapid Changes in Clonal Lines: The Death of a 'Sacred Cow'." *Biological Journal of the Linnean Society* 79: 3–16.

McLaughlin, B. and Bennett, K. (2005). "Supervenience." *Stanford Encyclopedia of Philosophy* <http://plato.stanford.edu/archives/fall2005/entries/supervenience/>.

McShea, D. W. (2002). "A Complexity Drain on Cells in the Evolution of Multicellularity." *Evolution* 56: 441–52.

Mameli, M. (2004). "Nongenetic Selection and Nongenetic Inheritance." *British Journal for the Philosophy of Science* 55: 35–71.

Margulis, L. (1970). *Origin of Eukaryotic Cells: Evidence and Research Implications for a Theory of the Origin and Evolution of Microbial, Plant, and Animal Cells on the Precambrian Earth*. New Haven, CT: Yale University Press.

Matessi, C. and Jayakar, S. D. (1976). "Conditions for the Evolution of Altruism under Darwinian Selection." *Theoretical Population Biology* 9: 360–87.

Maynard-Smith, J. (1976), "Group Selection." *Quarterly Review of Biology* 61: 277–83.

——— (1987). "Reply to Sober," in John Dupré (ed.), *The Latest on the Best: Essays on Evolution and Optimality*. Cambridge, MA: MIT Press, 147–50.

——— (1988). "Evolutionary Progress and Levels of Selection," in M. H. Nitecki (ed.), *Evolutionary Progress*. Chicago, IL: University of Chicago Press, 219–30.

——— (1998). "The Origin of Altruism." *Nature* 393: 639.

——— and Szathmáry, E. (1995). *The Major Transitions in Evolution*. Oxford: W. H. Freeman.

Mayr, E. (1976). "Typological versus Population Thinking," in *Evolution and the Diversity of Life*. Cambridge, MA: Harvard University Press, 26–9. Reprinted in E. Sober (ed.), *Conceptual Issues in Evolutionary Biology* (2nd edn.) Cambridge, MA: MIT Press, 1994, 157–60.

——— (1963). *Animal Species and Evolution*. Cambridge, MA: Harvard University Press.

Medin, D. and Atran, S. (eds.) (1999). *Folkbiology*. Cambridge, MA: MIT Press.

Mesoudi, A., Whiten, A., and Laland, K. (2004). "Perspective: Is Human Cultural Evolution Darwinian? Evidence Reviewed from the Perspective of the Origin of Species." *Evolution* 58: 1–11.

Michener, C. D. (1974). *The Social Behavior of the Bees: A Comparative Study*. Cambridge, MA: Harvard University Press.

Michod, R. E. (1999). *Darwinian Dynamics: Evolutionary Transitions in Fitness and Individuality*. Princeton, NJ: Princeton University Press.

_____ (2006). "The Group Covariance Effect and Fitness Trade-offs during Evolutionary Transitions." *Proceedings of the National Academy of Sciences (USA)* 103: 9113–17.

_____ and Roze, D. (2001). "Cooperation and Conflict in the Evolution of Multicellularity." *Heredity* 81: 1–7.

_____ and Sanderson, M. (1985). "Behavioural Structure and the Evolution of Social Behavior," in P. J. Greenwood and M. Slatkin (eds.), *Evolution: Essays in Honour of John Maynard Smith*. Cambridge: Cambridge University Press, 95–104.

_____ Nedelcu, A. M., and Roze, D. (2003). "Cooperation and Conflict in the Evolution of Individuality IV: Conflict Mediation and Evolvability in *Volvox carteri*." *BioSystems* 69: 95–114.

Mills, S. K. and Beatty, J. H. (1979). "The Propensity Interpretation of Fitness." *Philosophy of Science* 46: 263–86.

Millstein, R. (2002). "Are Random Drift and Natural Selection Conceptually Distinct?" *Biology and Philosophy* 17: 33–53.

_____ (2006). "Natural Selection as a Population-Level Causal Process." *British Journal for the Philosophy of Science* 57: 627–53.

Mitton, J. B. and Grant, M. C. (1996). "Genetic Variation and the Natural History of Quaking Aspen." *BioScience* 46: 25–31.

Molinier J., Ries G., Zipfel C., and Hohn B. (2006). "Transgeneration Memory of Stress in Plants." *Nature* 442: 1046–9.

Moss, L. (2003). *What Genes Can't Do*. Cambridge, MA: MIT Press.

Muller, H. J. (1932). "Some Genetic Aspects of Sex." *American Naturalist* 66: 118–38.

Nanay, B. (2002). "The Return of the Replicator: What is Philosophically Significant in a General Account of Replication and Selection? *Biology and Philosophy* 17: 109–21.

_____ (2005). "Can Cumulative Selection Explain Adaptation?" *Philosophy of Science* 72: 1099–112.

_____ (forthcoming). "Trope Nominalism and Anti-Essentialism about Biological Kinds."

Neander, K. (1995). "Pruning the Tree of Life." *British Journal for the Philosophy of Science* 46: 59–80.

Nowak, M. A. (2006). *Evolutionary Dynamics*. Cambridge, MA: Harvard University Press.

Nunney, L. (1985). "Group Selection, Altruism, and Structured-Deme Models." *American Naturalist* 126: 212–30.

O'Malley, M. and Dupré, J. (2007). "Size Doesn't Matter: Towards a More Inclusive Philosophy of Biology." *Biology and Philosophy* 22: 155–91.

Oborny, B. and Kun, A. (2002). "Fragmentation of Clones: How does it Influence Dispersal and Competitive Ability?" *Evolutionary Ecology* 15: 319–46.

Odling-Smee, F. J., Laland, K. N., and Feldman, M. W. (2003). *Niche Construction: The Neglected Process in Evolution*. Princeton, NJ: Princeton University Press.

Okasha, S. (2003a). "The Concept of Group Heritability." *Biology and Philosophy* 18: 445–61.

—— (2003b). "Does the Concept of 'Clade Selection' Make Sense?" *Philosophy of Science* 70: 739–51.

—— (2006). *Evolution and the Levels of Selection*. Oxford: Oxford University Press.

Otto, S. P. and Hastings, I. M. (1998). "Mutation and Selection Within the Individual," *Genetic,* 102/103: 507–24.

Oyama, S. (1985). *The Ontogeny of Information, Developmental Systems and Evolution*. Cambridge: Cambridge University Press.

Pearse, A-M. and Swift, K. (2006). "Allograft Theory: Transmission of Devil Facial-Tumour Disease." *Nature* 439: 549.

Pennock, R. T. (ed.), (2001). *Intelligent Design Creationism and Its Critics: Philosophical, Theological, and Scientific Perspectives*. Cambridge, MA: MIT Press.

Pigliucci, M. and Kaplan, J. (2006). *Making Sense of Evolution: The Conceptual Foundations of Evolutionary Biology*. Chicago, IL: University of Chicago Press.

Pineda-Krch, M. and Lehtilä, K. (2004). "Costs and Benefits of Genetic Heterogeneity within Organisms." *Journal of Evolutionary Biology* 17: 1167–77.

Poincaré, H. (1905/1952). *Science and Hypothesis*. New York: Dover.

Pollan, M. (2002). *The Botany of Desire: A Plant's-Eye View of the World*. New York, NY: Random House.

Potochnik, A. (2007). "Evolution, Explanation and Unity of Science." PhD dissertation, Stanford University.

Preston, K. and Ackerly. D. (2004). "The Evolution of Allometry in Modular Organisms," in M. Pigliucci and K. Preston (eds.), *Phenotypic Integration. Studying the Ecology and Evolution of Complex Phenotypes*. Oxford: Oxford University Press, 80–106.

Price, G. R. (1970). "Selection and Covariance." *Nature* 227: 520–1.

—— (1972). "Extension of Covariance Selection Mathematics." *Annals of Human Genetics* 35: 485–90.

—— (1995). "The Nature of Selection." *Journal of Theoretical Biology* 175: 389–96.

Prusiner, S. B. (1998). "Prions." *Proceedings of the National Academy of Sciences (USA)* 95: 13363–83.

Queller, D. C. (1985). "Kinship, Reciprocity and Synergism in the Evolution of Social Behaviour." *Nature* 318: 366–7.

—— (1997). "Cooperators Since Life Began (Review of *The Major Transitions in Evolution,* by J. Maynard Smith and E. Szathmáry)." *Quarterly Review of Biology* 72: 184–8.

_____ and Strassmann, J. E. (2002). "The Many Selves of Social Insects." *Science* 296: 311–13.

Reisman, K. (2005). "Conceptual Foundations of Cultural Evolution." PhD dissertation, Stanford University.

_____ and Forber, P. (2005). "Manipulation and the Causes of Evolution." *Philosophy of Science* 72: 1113–23.

Rice, S. H. (2004), *Evolutionary Theory: Mathematical and Conceptual Foundations*. Sunderland, MA: Sinauer.

Richerson, P. J. and Boyd, R. (2005). *Not by Genes Alone: How Culture Transformed Human Evolution*. Chicago, IL: University of Chicago Press.

Ridley, M. (1996). *Evolution* (2nd edn.). Cambridge, MA: Blackwell.

_____ (2000). *Mendel's Demon: Gene Justice and the Complexity of Life*. London: Weidenfeld & Nicolson.

Rinkevich, B. (2004). "Will Two Walk Together, Except They Have Agreed? Amos 3:3." *Journal of Evolutionary Biology* 17: 1178–9.

Rogers, D. and Ehrlich, P. (2008). "Natural Selection and Cultural Rates of Change." *Proceedings of the National Academy of Sciences (USA)* 105: 3416–20.

Rosenberg, A. (1994). *Instrumental Biology, Or the Disunity of Science*. Chicago, IL: University of Chicago Press.

Ross, C. N., French, J. A., and Orti, G. (2007). "Germ-line Chimerism and Paternal Care in Marmosets (*Callithrix kuhlii*)." *Proceedings of the National Academy of Sciences (USA)* 104: 6278–82.

Roughgarden, J. (1979). *Theory of Population Genetics and Evolutionary Ecology: An Introduction*. New York, NY: Macmillan.

Rutherford, S. (2000). "From Genotype to Phenotype: Buffering Mechanisms and the Storage of Genetic Information." *Bioessays* 22: 1095–105.

Sachs, J. L., Mueller, U. G., Wilcox, T. P., and Bull, J. J. (2004). "The Evolution of Cooperation." *Quarterly Review of Biology* 79: 135–60.

Salipante, S. J. and Horwitz, M. S. (2006). "Phylogenetic Fate Mapping." *Proceedings of the National Academy of Sciences (USA)* 103: 5448–53.

Santelices, B. (1999). "How Many Kinds of Individual are There?" *Trends in Ecology & Evolution,* 14: 152–5.

Schlichting, C. D. and Pigliucci, M. (1998). *Phenotypic Evolution: A Reaction Norm Perspective*. Sunderland, MA: Sinauer.

Schlosser, G. and Wagner, G. P. (eds.) (2004). *Modularity in Development and Evolution*. Chicago, IL: University of Chicago Press.

Scriven, M. (1959). "Explanation and Prediction in Evolutionary Theory." *Science* 130: 477–82.

Shtulman, A. (2006). "Qualitative Differences between Naïve and Scientific Theories of Evolution." *Cognitive Psychology* 52: 170–94.

Skyrms, B. (1994). "Darwin Meets The Logic of Decision." *Philosophy of Science* 61: 503–28.

——— (2003). *The Stag Hunt and the Evolution of Social Structure*. Cambridge: Cambridge University Press.

Sober, E. (1980). "Evolution, Population Thinking, and Essentialism." *Philosophy of Science*. 47: 350–83.

——— (1984). *The Nature of Selection: Evolutionary Theory in Philosophical Focus*. Cambridge, MA: MIT Press.

——— (1992). "The Evolution of Altruism: Correlation, Cost, and Benefit." *Biology and Philosophy* 7:177–88.

——— (1995). "Natural Selection and Distributive Explanation: A Reply to Neander." *British Journal for the Philosophy of Science* 46: 384–97.

——— and Lewontin, R. C. (1982). "Artifact, Cause and Genic Selection." *Philosophy of Science* 49: 157–80.

——— and Wilson, D. S. (1998). *Unto Others: The Evolution and Psychology of Unselfish Behavior*. Cambridge, MA: Harvard University Press.

Spencer, H. (1871). *Principles of Psychology* (2nd edn.) New York, NY: Appleton.

Sperber, D. (1996). *Explaining Culture: A Naturalistic Approach*. Oxford: Blackwell.

——— (2000). "An Objection to the Memetic Approach to Culture," in Robert Aunger (ed.), *Darwinizing Culture: The Status of Memetics as a Science*. Oxford and New York: Oxford University Press, 163–73.

Stegmann, U. (forthcoming). "Selection and the Explanation of Traits."

Stephens, C. (2004) "Selection, Drift, and the 'Forces' of Evolution." *Philosophy of Science* 71: 550–70.

Stephens, D. (1991). "Change, Regularity and Value in the Evolution of Animal Learning." *Behavioral Ecology* 2: 77–89.

Sterelny, K. (2001). "Niche Construction, Developmental Systems and the Extended Replicator," in R. Gray, P. Griffiths, and S. Oyama (eds.), *Cycles of Contingency: Developmental Systems and Evolution*. Cambridge, MA: MIT Press, 333–49.

——— (2003). *Thought in a Hostile World*. Malden, MA: Blackwell.

——— (2006). "Memes Revisited." *British Journal for Philosophy of Science* 57: 145–65.

——— (forthcoming). "The Evolution and Evolvability of Culture." *Mind and Language*.

——— and Calcott, B. (eds.) (forthcoming). *Major Transitions in Evolution Revisited*. Altenberg: Konrad Lorenz Institute.

_____ and Griffiths, P. E. (1999). *Sex and Death: An Introduction to the Philosophy of Biology*. Chicago, IL: University of Chicago Press.

_____ and Kitcher, P. (1988). "The Return of the Gene." *Journal of Philosophy* 85: 339–61.

_____ Smith, K., and Dickison, M. (1996). "The Extended Replicator." *Biology and Philosophy* 11: 377–403.

Strevens, M. (1998). "Inferring Probabilities From Symmetries." *Noûs* 32: 231–46.

Taylor, P. and Jonker, L. (1978). "Evolutionarily Stable Strategies and Game Dynamics." *Mathematical Biosciences* 40: 145–56.

Templeton, A. R. (1989). "The Meaning of Species and Speciation: A Genetic Perspective," in D. Otte and J. A. Endler (eds.), *Speciation and its Consequences*. Sunderland, MA: Sinauer, 3–27.

Thompson, M. (1995). "The Representation of Life," in R. Hursthouse, G. Lawrence, and W. Quinn (eds.), *Virtues and Reasons*. Oxford: Oxford University Press, 247–97.

Tomasello, M. (1999). *The Cultural Origins of Human Cognition*. Cambridge, MA: Harvard University Press.

Trivers, R. L. (1971). "The Evolution of Reciprocal Altruism." *Quarterly Review of Biology* 46: 35–57.

Tuomi, J. and Vuorisalo, T. (1989a). "What Are the Units of Selection in Modular Organisms?" *Oikos* 54: 227–33.

_____ _____ (1989b). "Hierarchical Selection in Modular Organisms." *Trends in Ecology and Evolution* 4: 209–13.

Turner, J. S. (2000), *The Extended Organism: The Physiology of Animal-Built Structures*. Cambridge, MA: Harvard University Press.

Uyenoyama, M. and Feldman, M. W. (1980). "Theories of Kin and Group Selection: A Population Genetics Perspective." *Theoretical Population Biology* 17: 380–414.

Vuorisalo, T. and Tuomi, J. (1986). "Unitary and Modular Organisms: Criteria for Ecological Division." *Oikos* 47: 382–5.

Wade, M. J. (1978). "A Critical Review of the Models of Group Selection." *Quarterly Review of Biology* 53: 101–14.

_____ (1985). "Soft Selection, Hard Selection, Kin Selection, and Group Selection." *American Naturalist* 125: 61–73.

Walsh, D. M., Lewens, T., and Ariew, A. (2002), "The Trials of Life: Natural Selection and Random Drift." *Philosophy of Science* 69: 452–73.

Weatherson, B. (2005). "Intrinsic vs. Extrinsic Properties." *Stanford Encylopedia of Philosophy*. <http://plato.stanford.edu/entries/intrinsic-extrinsic/>.

Weisberg, M. (2007). "Who is a Modeler?" *British Journal for Philosophy of Science* 58: 207–33.

Weismann, A. (1896). *On Germinal Selection.* (Trans. by T. McCormack). Chicago, IL: Open Court.

Weismann, A. (1909). "The Selection Theory," in A. C. Seward (ed.), *Darwin and Modern Science*: Cambridge: Cambridge University Press, 23–86.

West-Eberhard, M. J. (2003). *Developmental Plasticity and Evolution.* New York: Oxford University Press.

White, J. (1979). "The Plant as a Metapopulation." *Annual Review of Ecology and Systematics* 10: 109–45.

Whitham, T. G. and Slobodchikoff, C. N. (1981). "Evolution by Individuals, Plant–Herbivore Interactions, and Mosaics of Genetic Variability: The Adaptive Significance of Somatic Mutations in Plants." *Oecologia* 49: 287–92.

Wilkins, J. S. (2003). "How to be a Chaste Species Pluralist-Realist: The Origins of Species Modes and the Synapomorphic Species Concept." *Biology and Philosophy* 18: 621–38.

Williams, G. C. (1966), *Adaptation and Natural Selection: A Critique of some Current Evolutionary Thought.* Berkeley, CA: University of California Press.

—— (1992). *Natural Selection: Domains, Levels, and Challenges.* New York, NY: Oxford University Press.

Wilson, D. S. (1975). "A Theory of Group Selection." *Proceedings of the National Academy of Sciences* 72: 143–6.

—— (1980). *The Natural Selection of Populations and Communities.* Menlo Park, CA: Benjamin Cummings.

Wilson, E. O. (1971). *The Insect Societies.* Cambridge, MA: Harvard University Press.

Wilson, R. A. (2005). *Genes and the Agents of Life: The Individual in the Fragile Sciences.* Cambridge: Cambridge University Press.

—— (2007). "The Biological Notion of Individual." *Stanford Encyclopedia of Philosophy.* <http://plato.stanford.edu/archives/fall2007/entries/biology-individual/>.

Wimsatt, W. (1980). "Reductionist Research Strategies and their Biases in the Units of Selection Controversy," in T. Nickles (ed.), *Scientific Discovery: Case Studies.* Dordrecht: Reidel, 213–59.

Winther, R. G. (2006). "Fisherian and Wrightian Perspectives in Evolutionary Genetics and Model-Mediated Imposition of Theoretical Assumptions." *Journal of Theoretical Biology,* 240: 218–32.

Woese, C. (2002). "On the Evolution of Cells." *Proceedings of the National Academy of Sciences* (*USA*) 99: 8742–7.

Wolpert, L. and Szathmáry, E. (2002). "Multicellularity: Evolution and the Egg." *Nature* 420: 745.

Wright, S. (1932). "The Roles of Mutation, Inbreeding, Crossbreeding and Selection in Evolution." *Proceedings of the Sixth International Congress of Genetics* 1: 257–66.

Wynne-Edwards, V. C. (1962). *Animal Dispersion in Relation to Social Behavior.* Edinburgh: Oliver & Boyd.

INDEX

abstraction:
  and organization of evolutionary
      theory 30–1
  and summaries of evolution by natural
      selection 17–20, 31
Adelson, G 15
agential view of evolution 5, 10–11, 142–3,
      144–5, 160
  and replicator approach 10, 36–7
  and teleological outlook on biological
      activity 12–13
age-structured populations 22, 23, 26
algae 73, 95–7, 109
altruism:
  and correlated interaction 118, 120–1,
      174–6
  and group selection 115–17, 118–19
  and kin selection 115, 119–20, 176–8
  and lattice population structure 117–18
  and reciprocity 115, 120
aphids 72
apomixis 71, 72, 106–7
Ariew, A 72
Aristotle 12, 79, 143
artifacts, as cultural replicators 154–5
asexual reproduction 33, 35, 36–7, 56, 66,
      72–3, 81–2, 89, 96, 104, 106–7, 127
aspen trees (*Populus tremuloides*) 71, 72,
      76, 81, 92, 95

bacteria, and genes 136, 137, 140
Beatty, J 60
bees 93, 97–9
bird nests 90, 155
bottlenecks: 94–100
  and collective entities/reproducers 91–2,
      94–6, 123–7
  and evolutionary transitions 123, 124
  and germ line 127
  and origin explanations 91–2
  and simple reproducers 100

and suppression of lower level
      evolution 101–2
  and variation 91–2, 102, 127
Brandon, R 29, 169, 171
Burt, A 133, 135, 139
Buss, L 122, 124 n5, 127

Calcott, B 124
causation:
  and cultural evolution 153–4
  and fitness/drift relation 27–8
  and levels/units of selection 111–13
cells:
  and compared with genes 130–1
  as Darwinian individuals 56, 113–14,
      126
  and evolution of multicellular
      organisms 101–3
  and intra-cell conflict 140–1
  and simple reproducers 90–1
Charlesworth, B 23, 30
Chartier, É-A 155
chimeras 75–76, 78
chloroplasts 74
chromosomes 88, 89, 130, 133–36, 138,
      140–41
cistrons 136, 137
clades, and reproduction 105
classical approach to evolution by natural
      selection 4–5, 19–20, 32
  and compared with replicator
      approach 32–3
clonality, and reproduction 71–2, 106
Coates, A G 72
collective entities/reproducers: 8, 9, 70,
      73–75, 87–91, 124–5
  and bottlenecks 124–5
  and germ line 92–3, 94
  and reproduction 70, 73–5, 87, 89,
      91–100, 101–102
colonies 73

competition:
  and Darwinian populations 9, 48, 51, 52, 125
  and genes 131, 133
  and origin explanations 51, 52
  and population growth rate 52
  and weak and strong senses 51
contextual analysis, and levels of selection 112 n2
continuity 57–9, 60–1, 63–6
Cook, R 71, 72 n1
cooperation:
  and levels of selection 115–16
  and social learning 156
  see also altruism
corals 71, 73, 74, 75, 79, 105
correlated interaction 118, 120–1, 174–8
cultural evolution 9, 65, 147–8, 149–50, 162–4, 162, 163–4
  and artifacts 154–5
  and cooperation 156
  and reproduction 150–1, 152–3, 153–5, 156–60, 163
  and levels/units of selection 162
  and population thinking 148–9
  and replicators 151, 160–1

dandelions 71, 81
Darwin, Charles: 1, 17, 19, 20, 24, 31, 107, 139, 164
Darwinian individuals 6, 40, 85–6, 109
  and genes 130, 136, 145
Darwinian populations: 6, 8, 9, 42, 107–8, 110
  and boundaries of 48–9, 53, 139–40
  and competition 48–53, 52, 125
  and distribution explanations 42, 43
  and evolutionary transitions 8–9, 122–24
  and levels/units of selection 7–8, 110, 129–35
  and marginal cases 6–7, 41, 103–7
  and minimal concept of 6, 7, 39–40, 41
  and origin explanations 42–3
  and paradigm cases 6, 7, 9, 41, 48–9, 63–7, 100–1
Dawkins, R:
  and cultural evolution 151, 160, 162
  and gene's eye view of evolution 129, 139, 142–4, 144 n7,
  and replicators 5, 32–3, 35 n9, 36–7, 47, 83
  and reproduction 90, 105, 105 n9, 106
de-Darwinization 100–103
Dennett, D C 32, 26, 161–2
development, and reproduction 84, 100

Dinka tribe 151–2
distribution explanations 42, 43
DNA 47, 130, 131–3, 135
Downes, S 169
drift 27–28, 31, 51, 55, 59–62

Ehrlich, P 155
Ender, J 173–4
error catastrophe 44, 45
essentialist causal model of organisms 12
eukaryotic cells 8, 74–5, 122, 124, 140
eusocial insects 93, 97–9, 123
evolution by natural selection:
  and abstract approach to 1–2
  and creative role of 49–50
  and Darwinian populations 107–8
  and distribution explanations 42
  and fitness/drift distinction 27–8, 59–62
  and heredity 24–6
  and origin explanations 42–3, 106–7
  and population thinking 11–12, 13
  and reproduction 40, 104
  and summaries of 17–20, 26–7, 31, 42, 173–4
  and teleological thinking 12–13
  and tree of life 1, 14–15, 108, 164
evolutionary nominalism 35, 168, 177
evolvability 41
extrinsic properties 53–6, 60
eye, human 50–1

facilitation 65
ferns 92, 78
fission, and reproduction 85
fitness:
  and drift 27–8, 59–62
  and human cells 56
  and intrinsic character 54–6
  and propensity view of 29–30
  and remote descendants 23
  and summaries of evolution by natural selection 21–4
fitness landscape 57–9, 61
folk psychology 13
folkbiology 12
Forber P 42 n1, 60
Francis, R 10, 144
fungi 73–4, 109

Gavrilets, S 58
generations, and alternation of 78–9
genes:
  and agential approach to evolution 142–5
  and competition 131, 133

and crossing-over 136, 137, 138, 140–1
as Darwinian individuals 129–35, 136, 145
and evolution of 135–42
and homing endonuclease genes 133–4
and intra-cell conflict 140–1
and meiotic drive 134
and particulate thinking 145
and replicator approach 5, 32–3, 129
and self-interest 133, 135, 142
and selfish genetic elements 133–4, 139
and transposons 131, 133, 134
as units of selection 129–30, 132–3
genets 71, 72, 105–6
germ line:
    and bottlenecks 127
    and collective entities/reproducers 92–3, 94–100, 106, 125
    and evolutionary transitions 123
    and plants 127
    and simple reproducers 100
    and suppression of lower-level evolution 102, 128
gestalt switching 115
Gonium 95–7
Gould, S J 17 n1, 43, 50, 85, 168
Grafen, A 140–1, 167, 168
Griesemer, J 81, 83–4, 100, 119 n3, 154
Griffiths, P 12, 42, 100, 106, 137 n3, 139
group selection, and altruism 115–17, 118–19
growth, and reproduction 70–2, 81–2, 105

Haig, D 10, 32, 36, 75, 76 n3, 140–1, 171
Hamilton, W 115, 119–20, 175–6, 178
Harper, J L 71, 72, 105 n9
heat-shock proteins 58–9
heredity 44–7, 63–6
    and summaries of evolution by natural selection 24–6
heritability 24–5, 168–172
hierarchies 109
homing endonuclease genes 133–4
Hull, D 5, 13, 83, 90, 31–32,
humans:
    and chimerism 76
    and levels of Darwinian processes 114, 136–7
hypercycle model 79

idealization:
    and models of evolution 6, 22, 24, 25, 26, 31
    and analysis of reproduction 82, 89

identity, and genetic criteria for 81–2
imitation rules, and cultural evolution 156–60, 163
individuality 69, 70, 84–6
inheritance, *see* heredity; reliability of inheritance
integration 91, 93–6, 100, 105
interactors, and replicator approach 32
intraorganismal genetic heterogeneity 75
intrinsic character 54–6, 63–6, 102–3, 125–6

Jackson, J B 72
Janzen, D H 71–2, 81, 105
jellyfish 79
Jenkin, F 45–6

kin selection, and altruism 115, 119–20, 176–8
Kerr, B 115, 123, 175, 178
Kuhn, T 2

landscapes, and fitness landscape 57–9
language 47, 163
Lennox, J 79 n6
levels of selection:
    and altruism 115–16
        correlated interaction 118, 120–1, 174–6
        group selection 115–17, 118–19, 174–6
        kin selection 115, 119–20, 176–8
        lattice population structure 117–18, 175–6
    and biological hierarchies 109–11
    and causation 111–13
    and cultural evolution 162
    and evolutionary transitions 122–4
    and gestalt switching 115
    and humans 113–14
    and modules 110
    and part/whole relations 101, 109
    and replicator approach 110–11
    and contextual analysis 112 n2
    and Price equation 112 n2
    and subversion of higher level evolution 101
    and suppression of lower level evolution 101–2
Lewontin, R C 18, 24, 47–8, 57, 65, 72, 110, 167, 172, 182
lichen 73–4, 109
LINE transposons 80–1, 133, 152
Lloyd, E A 36 n11, 111

McShea, D 126
marmosets 75
material overlap, and reproduction 83–4, 100
Maynard Smith, J 32, 85, 89–90, 122
Mayr, E 11–12, 13, 148
meiotic drive 134
memes 160–1
metamorphosis 104, 106
Michod, R E 122, 124 n3,
mitochondria 74
mosaics 75, 76–7, 82
   and Pinot meunier grape 77
multicellularity 8, 101–3, 122, 123, 126
multiplication, and reproduction 104, 106
mutation 28, 45, 49–51, 55, 59, 77, 82 n9, 91, 107

Nanay, B 51 n5
natural state model 12
Neander, K 42 n1
niche construction 63
nucleic acids, as replicators 47
Nuer tribe 151–2

oak trees 72, 76, 96, 110
Okasha, S 105, 111, 112 n2, 119, 130, 167, 171
origin explanations 42–3, 48–53, 91–2, 106

paranoia, Darwinian 10, 144–5, 161
particulate inheritance 45
persistence of entities 37–8 104, 105, 135–6, 151–2
philosophy of nature 3
philosophy of science 2–4
physiological individuals 71–2
Pinot meunier grape 77–8
plants:
   and germ line 127–8
   and mosaicism 76–7
Plato 11
Poincaré, H 62
population growth rate 22, 51–2
population thinking 11–12, 13, 148–9
Portuguese Man O' War 73, 79
Price equation 112 n2, 167, 171, 177, 180–1
prions 80, 152, 153
propensity view of fitness 29–30

quasi-independence of traits 47–8, 57
Queller, D 120, 124, 176–78

ramets 71, 76, 77, 81, 85, 107
rational choice 161
reciprocity, and altruism 115, 120
recombination, as source of variation 49
'Red Delicious' apple 71
Reisman, K 50 n4, 149
reliability of inheritance 44–7, 63–6
replicator dynamics 38, 165–6
replicator view of evolution 1–2, 5, 31–6, 38, 47, 129, 151
   and agential approach to evolution 10, 36–7
   and interactors 32
   and levels/units of selection 110–11
   and reproduction 35–6, 89–90
reproduction 69–70, 78–79, 83–6, 87–9, 94, 105–6
   and causal relations 79–81
   and collective entities/reproducers 70, 73–5, 87, 89, 91–100
      bees 97–9
      bottlenecks 91–2, 94
      colonies 73
      germ lines 92–3, 94
      integration 93–4
      spatial framework 94–100
      Volvox 95–7
   and cultural evolution 151–5
   and formal reproduction 79–81, 152
   and genetic criteria for identity 81–2
   and Griesemer's view of 83–4, 100, 154
   and growth 70, 71–2, 81–2
   and individuality 84–6
   and marginal cases 7, 103–7
   and metamorphosis 104, 106
   and mosaicism 75, 76–7, 82
   and multiplication 104, 106
   and scaffolded reproducers 88, 90
   and simple reproducers 87–8, 89, 90–1, 100
   and species selection 105
retroviruses 80, 84, 152
Ridley, M 19
Rogers, D 155
Rosenberg, A 60

Santelices, B 85
scaffolded reproducers 88, 90, 130, 100, 152, 154–5
Scriven, M 60 n9
sea anemones 71, 79
selfishness:
   and genes 133, 135, 139, 142
   and memes 160

sexual reproduction 44, 45–6, 49, 52, 66, 126–7
simple reproducers 87–8, 89, 90–1, 100
Skyrms, B 156, 157
Sober, E 12, 28, 59, 117
social learning 147, 155–60, 163
spatial representation:
  of collective reproducers 94–100
  of Darwinian populations 7, 8, 43–4, 63–7
species selection 105
Spencer, H 18
Stephens, C 59
Sterelny, K 36, 41, 88 n1, 90 n2, 106, 130, 139, 154
strawberry 71, 85
Strevens, M 62
struggle for life, and evolution 48
supervenience 54–5
symbiosis 73–5, 109
Szathmáry, E 32, 89–90, 122, 127

teleology 12–13
Templeton, A R 52 n6
Tomasello, M 163
transitions in evolution 8–9, 122–4

transitions in individuality 122
transposons 80–1, 131, 133, 134, 152
tree of life 1, 14–15, 108, 164
Trivers, R 133, 135, 139
types 34–5, 35 n9
typological thinking 11, 12

variation:
  and bottlenecks 91–2, 102, 127
  and origin explanations 42–3
  and population thinking 11, 12, 14, 16
  and quasi-independence of traits 57
  and reliability of inheritance 47
  and supply of 47–50
vehicles, and replicator approach 32
violets 71, 72
viruses, and reproduction 88
Volvox 95–7

Wade, M J 119 n3
Weismann, A 17–18, 24
Williams, G C 138
Wilson, D S 115, 117
Wilson, R 11, 85
Wright, S 57
Wynne-Edwards, V C 142